Aquaculture and
the Environment

Aquaculture and the Environment

Second edition

T.V.R. Pillay

Former Programme Director
Aquaculture Development and Coordination Programme
Food and Agriculture Organization of the United Nations
Rome, Italy

Blackwell
Publishing

© 1992, 2004 by Fishing News Books, an imprint of Blackwell Publishing Ltd

Editorial Offices:
Blackwell Publishing Ltd, 9600 Garsington Road, Oxford OX4 2DQ, UK
  Tel: +44 (0)1865 776868
Blackwell Publishing Professional, 2121 State Avenue, Ames, Iowa 50014-8300, USA
  Tel: +1 515 292 0140
Blackwell Publishing Asia Pty Ltd, 550 Swanston Street, Carlton, Victoria 3053, Australia
  Tel: +61 (0)3 8359 1011

First published 1992 by Fishing News Books, an imprint of Blackwell Scientific Publications
Second edition published 2004

Library of Congress Cataloging-in-Publication Data
Pillay, T.V.R.
    Aquaculture and the environment / T.V.R. Pillay.—2nd ed.
      p. cm.
    Includes bibliographical references (p.).
    ISBN 1-4051-0167-9 (alk. paper)
      1. Aquaculture. 2. Aquaculture–Environmental aspects. 3. Aquatic animals–
Effect of water pollution on. 4. Environmental impact analysis. I. Title.

SH135.P55 2003
639.8–dc21   2003045325

ISBN 1-4051-0167-9

A catalogue record for this title is available from the British Library

Set in 10/12pt Melior
by Graphicraft Limited, Hong Kong

For further information on Fishing News Books, visit our website:
www.fishknowledge.com

Contents

Preface

Modern aquaculture is a new science, though traditional in origin, and the industry based on it is of recent development. Nevertheless, aquaculture production has steadily increased from about 5 million tons in 1973 to 39.4 million tonnes in 1998. This production comes from a variety of farms ranging from small-scale owner-operated fish ponds to larger-scale co-operative and corporate farms, supported by auxiliary industries such as feed and equipment manufacture.

Increased production is being achieved by the expansion of areas of land and water under culture, and the use of more intensive and modern farming technologies that involve higher usage of inputs such as water, feeds, fertilizers and chemicals. Higher inputs normally give rise to increased waste discharges from farms, and extension of cultivated areas, whether on land or water, gives rise to conflicts with other users of the resources.

Socio-cultural and economic impacts of some of the large-scale projects have not always benefited the disadvantaged local populations, although this was one of the major aims of development, especially in developing countries.

As a result aquaculture, which was once considered an environmentally sound practice because of its traditional polyculture and integrated systems of farming based on optimum utilization of farm resources, including farm wastes, is now counted among the potential polluters of the aquatic environment and the cause of degradation of wetland areas.

More and more restrictions are being imposed on aquaculture ventures in industrially advanced, as well as in developing, countries. Environmental impact assessments (EIAs) are becoming essential in most countries to obtain necessary permission from designated authorities to start farming projects, except for very small-scale operations where no adverse impacts are expected.

Aquaculture research, which is still in its infancy, has devoted much of its attention so far to improving existing technologies or developing new ones for increased production. Environmental studies have been limited mostly to determining optimum conditions required in the farms and not to evaluating the impact of farming on the external environment.

As a result of problems that have arisen from pen and cage farming in marine or enclosed freshwater areas, and the increasing worldwide awareness of environmental consequences of developmental activities in general, efforts are now under way to obtain some of the basic information needed to make appropriate environmental impact assessments of at least some of the culture systems, and to design sound management strategies. Information already available or deducible from analogous situations has to be the basis for predicting the possible impact of new development projects. Access to existing information and experience is therefore of prime importance in responding to the requirements of obtaining permits for initiating farming ventures.

This book is an attempt to bring together available information pertinent to environmental consequences of aquaculture development, with a view to assisting in the evaluation of future projects. Results of investigations so far seem to indicate that the adverse impact of properly organized aquaculture on the environment is minimal, and most of the potential adverse changes are not irreversible and can be avoided or mitigated by appropriate measures.

Though many of the potential environmental concerns are of a speculative nature, and are not borne out by scientific evidence, the paucity of adequate research makes any definitive judgement hazardous. Reviews of existing information contained in the book focus on the many gaps that need to be filled through interdisciplinary research.

Many ecological disasters that have occurred as a result of unsustainable use, abuse and misuse of natural resources have clearly demonstrated that long-term and sustainable development can be achieved only through sound environmental management. Experience in aquaculture to date strongly reinforces this concept, and aquaculturists are becoming more and more aware of the need for a long-term perspective in development planning, and the incorporation of solutions to the socio-cultural and aesthetic concerns of affected communities, in site selection, farm design and operation.

The book therefore includes suggested procedures for the environmental management of aquaculture comprising the major elements of planning, information, impact assessments, and mitigatory measures supported by technological improvements, increased research and legislation.

Environmental impact assessment, which is an essential tool in environmental management, is a relatively new and growing technology, where problems are constantly encountered and pragmatic solutions are sought for predicting and mitigating impacts. The validity of the predictions has to be verified by appropriate monitoring and reassessment of the mitigatory measures.

In view of this, it is pointless to insist on rigid procedures for the initial assessments, especially in the case of aquaculture where the impact is generally low and localized. Relatively simple methods of assessment and evaluation of alternative mitigation measures that appear to be suitable for aquaculture projects are described in the book.

The concept of integrating impact assessment (preliminary or detailed as required) with all the phases of a project, from feasibility studies to project design and operation, along with impact monitoring/auditing for sustainable development, is described. The steps defined for such integration, together with data on environmental impacts of various aquaculture systems in different ecological conditions are expected to assist aqua-culturists and investors to formulate and implement environmentally acceptable and economically viable projects. Most of the current information comes from aquaculture in farms but, as a result of dimin-ishing capture-fishery production, stock enhancement and culture-based fisheries are becoming important. Environmental consequences of these forms of development are addressed and guidelines on selection of water bodies and the stocks to be enhanced are reviewed.

In attempting to cover all the major types of impact of known systems of aquaculture and stock enhancement on a worldwide basis, it is likely that some of less or local importance have been overlooked. Further, no distinction has been made in the treatment of environmental problems in industrially advanced and developing countries, as the basic concepts, in so far as aquaculture is concerned, are not very dissimilar, even though legal restrictions are often less stringent, or even non-existent, in some countries of the latter group.

T.V.R. Pillay

Acknowledgements

I have received help from a number of people in the preparation of this book. I would specially like to acknowledge the assistance of Dr. Michael J. Phillips (NACA, Bangkok), Mr. Heiner C.F. Naeve (FAO, Rome), Mr. Imre Csavas (Hungary), Ms. I. Bertilsson (FAO, Rome), Dr. Henri Grizzel (IFREMER, France), Dr. Kee Chai Chong (BOBP, Chennai) and Mr. Michael New (UK).

Ms. Bertilsson read through the manuscript of the chapter on environmental management of aquaculture and offered helpful suggestions. My wife, Dr. Sarojini Pillay, gave valuable assistance in the finalization of the manuscript. Ms. Rhona Wood typed the manuscript of the first edition with characteristic care and patience during her spare time and I am grateful for her timely assistance. It is a pleasure to record my appreciation of the help of Mr. Sujay Jayakumaran in the final preparation of this revision.

Chapter 1
Introduction

All forms of food production, like any other human activity, affect the environment in one way or another. Some of these may be considered beneficial, while others are not consistent with long-term preservation of natural ecosystems. Disturbance to the balance of nature is a recognized phenomenon, but as long as the pressures on the environment remain within sustainable limits that permit continuing natural adjustment, no major conflicts are recognized.

The combination of population increases and major multiple demands on natural resources as a result of agricultural and industrial development in the recent past has focused world attention on the need to retain life-support systems and the amenities of land, water and air for the enjoyment and wellbeing of future generations. The degree and intensity of concern are generally in direct proportion to the extent of exploitation of resources and industrial development. This has obviously contributed to the emergence of environmental concern as a major socio-economic and political force in the industrialized world. However, environmental effects do not conform to political or geographic boundaries and, therefore, the concept of a global environment is gaining ground.

The need to learn from the experience of ecologically unsustainable development and to prevent its repetition on a global basis has been generally accepted, even though in the developing parts of the world there is a conflict of priorities. The need for rapid economic development, increasing food production for feeding the fast-expanding human population and developing physical infrastructure and amenities to improve the standard of living all have very high priority in developing countries.

In the complex and conflicting situation in which resource management decisions have to be made, neither complete destruction of the natural environment nor complete avoidance of resource exploitation can be practical. A logical course would, therefore, be a balance between rational use, conservation and preservation in order to optimize man's use of natural resources on a long-term basis.

In the global context of the environmental impact of human activities and interference, such as agriculture, habitation, industry, infrastructure development etc., the contribution of aquaculture is undoubtedly small.

Because of the stringent water quality and environmental requirements of aquatic farms, aquaculturists seldom recognized the possibility
of aquaculture being considered a polluter of the environment. In the
type of farming that was being practised, quality of water released from
the farms was often much better than that of the inflow from natural
sources. Many traditional fish culture systems functioned as efficient
means of recycling agricultural and domestic wastes, and thus contributed
to the abatement of environmental pollution.

Some concern existed among public health authorities in certain tropical countries, especially in Africa, that stagnant fish ponds may serve
as breeding grounds for mosquitoes that transmit malaria or harbour
vectors of schistosomiasis. This possibility could be avoided through
proper design and management of the ponds.

Another concern was about the impact of collecting wild eggs, larvae
or fry of species for culture. Fishermen often ascribed lower catches from
wild stocks of these species to gathering of the early stages for rearing in
aquaculture facilities. Scientific investigations have not yet produced any
conclusive proof in favour of this belief, as the eggs and larvae collected
from the wild were usually a small proportion of those that would
normally die under natural conditions.

Nevertheless, aquaculturists all over the world have consistently tried
to produce the necessary seed stock under controlled conditions in
hatcheries, and not to depend on natural sources. In addition, hatchery-
produced young were used in many areas for the enhancement of wild
stocks. A number of transplantations and introductions of exotic species
for culture have been made, and many of them have become established
in the new environments with very few adverse effects, providing important sources of food or recreation. But indiscriminate introductions that
may lead to environmental problems are discouraged and many governments have imposed restrictions on imports.

Thus, until a couple of decades ago, when technologies remained less
intensive, aquaculture was generally considered an environmentally
sound activity. A major change in perception has occurred in recent
years with the adoption of more intensive techniques of production and
the extension of farming to coastal and foreshore areas of the sea. The
current surge of worldwide concern with the effects of misguided development and the disastrous global environmental changes that are fore-
cast has forced developmental agencies and governments to reassess
their policies. Most of them now insist on an environmental assessment
as a prerequisite for all developmental programmes in order to ensure
their environmental soundness and sustainability, and to incorporate
means of minimizing, mitigating or compensating for adverse impacts,
if any.

Aquaculturists may wonder why large-scale aquaculture projects are
considered on a par with other well-known polluters that have brought
about irreversible changes to the environment. Besides the adoption of
intensive production technologies involving greater inputs and waste

discharges, two major reasons appear to be the rapid expansion of cage farming in coastal areas and the extensive reclamation of coastal wetlands for shrimp farming in tropical areas. Indiscriminate destruction of mangrove swamps has occurred in many countries, not only for aquaculture but also for other purposes, including property development, harbour and road construction, logging, salt-making, oil drilling, dumping of minetailings etc.

Cage farming in protected bays, fjords and other coastal areas involves the use of highly intensive stocking and feeding, which results in increased sedimentation, biochemical oxygen demand and nutrient loading. The crowding of fish pens and cages in semi-enclosed areas has sometimes resulted in autopollution and transmission of diseases.

There have also been conflicts with other uses of coastal areas, such as recreation, navigation and enjoyment of scenic beauty of waterfront areas, forcing regulatory agencies to control the expansion of these types of aquaculture.

It is obvious that a sustainable use of the environment has to be one that will not bring about irreparable harm to the ecosystem, but restrict the inevitable changes within limits of natural fluctuation. Appropriate control has to be enforced to ensure such rational use, but to be effective, guidelines and regulations have to be based on adequate knowledge of the impacts of developmental alternatives and possible means of mitigating the adverse effects of selected uses.

In actual practice, however, arbitrary restrictions are often imposed by extending regulations relating to agriculture, animal husbandry or industrial development to cover aquaculture, even though they are clearly inapplicable.

Though an ancient practice, aquaculture has a weak scientific base, and much of the research so far has been directed to the development of production technologies and the management of environmental conditions within aquaculture farms. However, as a result of public concern over the environmental impact of coastal farming, a number of studies have been undertaken, especially in Europe, to determine the source and nature of effluents discharged from fish farms and their fate in the receiving waterways.

Extensive discussions and assessments of the effects of reclaiming mangrove swamps have also been undertaken. Besides these, the environmental impacts of other aquaculture practices, such as the introduction of exotic species for stock enhancement, use of chemicals and water use patterns, have also received some scientific attention.

An effort is made in the following chapters to bring together the available information to facilitate environmental impact assessments and development actions and to focus attention on the need for critical studies on the impact of all types of aquaculture system in relevant geographic and climatic conditions.

Chapter 2
Water Quality

2.1 Aquaculture Farms

The major focus of attention in this book is the impact of aquaculture on the outside environment. Nevertheless, it is relevant to consider here the available information on minimum water quality levels that an aquatic farm has to maintain. As is obvious, the tolerance limits of water quality depend very much on the species cultivated, especially in respect of temperature and salinity. So we may restrict this review to only the water quality requirements that have significance in assessing environmental impact. These are mainly dissolved oxygen, pH, carbon dioxide, ammonia, nitrites, nitrates, hydrogen sulphide, pesticides and turbidity. The optimum levels of many of these are not accurately known for many species, but based on long-term sublethal toxicity tests and experience in experimental and production farms, 'safety levels' have been indicated (Tiews, 1981).

In salmonid and warm-water crustacean culture, dissolved oxygen levels are not allowed to go below 5 mg l^{-1} for more than a few hours. Eel, carp and tilapia in farms can tolerate lower concentrations ranging from 3 to 4 mg l^{-1}. The optimum levels may be higher than this. According to Swingle (1969) and Boyd (1981), warm-water fish survive at dissolved oxygen levels as low as 1 mg l^{-1}, but their growth is slowed down by prolonged exposure. The desirable range is above 5 mg l^{-1}.

Slow growth results from pH levels below 6–6.5, and the acid death point is reported to be pH 4. The alkaline death point is pH 11, but the desirable range for fish production is 6.5–9.0 at daybreak (Boyd, 1981).

In water used for intensive fish culture, free carbon dioxide levels fluctuate typically from 0 mg l^{-1} in the afternoon to 5–10 mg l^{-1} at daybreak with no obvious ill-effects on fish (Parks *et al.*, 1975). Higher concentrations of free carbon dioxide, even up to 60 mg l^{-1}, may be tolerated provided that dissolved oxygen concentrations are high.

Un-ionized ammonia (NH$_3$) is toxic to fish, but the ammonium ion (NH$_4^+$) is not toxic (Downing & Merkens, 1955; Boyd, 1981). According to EIFAC (1973), toxic levels of NH$_3$ for short-term exposure usually lie between 0.6 and 2.0 mg l^{-1}. Others consider the maximum tolerable concentration to be 0.1 mg l^{-1} (Tiews, 1981), the preferred level being below this. Un-ionized ammonia becomes more toxic in low concentrations of dissolved oxygen, but this is of little importance in pond farms as the

4

toxicity decreases with increasing carbon dioxide concentration, which is usually the case when the dissolved oxygen is high.

Available information on safety limits of nitrites (NO_2) is very limited, although studies indicate that nitrite may be a significant factor in channel catfish ponds. The suggested maximum level for prolonged exposure in hard fresh water is 0.1 mg l^{-1}. Suggested nitrate (NO_3) levels are below 100 mg l^{-1}.

Un-ionized hydrogen sulphide (H_2S) is extremely toxic to fish at concentrations that may occur in natural waters as well as in aquaculture farms. It has been demonstrated that concentrations of H_2S could result in poor growth of channel catfish (Bonn & Follis, 1967). Bioassays of several species of fish suggest that any detectable concentration of H_2S should be considered detrimental to fish production (Boyd, 1981).

Many pesticides, particularly insecticides, are extremely toxic to fish. Acute toxicity values for several commonly used insecticides range from 5 to 100 mg l^{-1} (Cope, 1964), and on longer exposure even lower concentrations may be toxic. Even when they do not cause mortality of the species under culture, they may affect the growth of food organisms and thus reduce their growth and productivity. Aquaculture farms, therefore, take all possible precautions to prevent pesticide contamination. If the farms have to be treated with pesticides to eradicate pests or predators, care is taken to protect the stock from exposure to the pesticides. Generally, the stock is released back into rearing units only after the water is no longer toxic.

Turbidity is also an important water quality which a fish farmer has to control. Turbidity caused by suspended soil particles has usually no direct effect on fish and shellfish, but it restricts light penetration and limits photosynthesis. Also, sedimentation of soil particles can destroy benthic communities and smother fish eggs. Turbidity caused by plankton is not harmful to fish, but clay turbidity exceeding 20 000 mg l^{-1} causes behavioural reactions in many species of fish. Appreciable mortality occurs at turbidities above 175 000 mg l^{-1}.

2.2 Open Waters for Stock Building and Stock Enhancement

Stock enhancement can be considered as an extension of aquatic farming as far as water quality is concerned, because the water quality requirements are very similar in different stages of its operation, particularly when hatchery-reared young are utilized for stocking purposes. Stock enhancement is undertaken to increase production from lakes and reservoirs, built by human intervention in the form of building dams across rivers, and streams for irrigation or hydro-electric power generation. Many coastal areas are affected by habitat degradation due to pollution, as well as over fishing, and need stock enhancement to upgrade production from commercial and recreational fishing (Ungson *et al.*, 1993; Isaksson, 1998). In most cases, young are produced and reared in hatcheries and the food resources of water bodies are increased by application of fertilizers or introduction of food organisms, along with the removal of predators and competitors. Since aquaculture-based fisheries are of different species, water quality and habitat improvement vary considerably.

More than 120 species have been used for stock-enhancement experiments. Some of these have been successful and already contribute to commercial or recreational fisheries (FAO, 1999). Examples are stocks of salmon, sturgeon, sea bream, flounder, tilapia, Chinese and Indian carp, coregonid, penaeid shrimp, scallops and oysters.

Environmental surveys that precede stock enhancement decisions include water temperature, salinity, transparency, pH, BOD, dissolved oxygen, nutrients such as phosphates, nitrates, silicates, suspended substances and soil conditions. Water-flow pattern in relation to dissipation time of waste material discharge from the hatcheries, or enhancement of the stocks have to be determined (Aure & Stigebrandt, 1990). Wind, waves, swells, tidal flows and currents, and the extent of shelf area around coasts also have to be evaluated with as much precision as possible.

As described in Chapter 5.4, categorization of types of stock enhancement is complicated by the variety of circumstances under which stocking is resorted to. However, the types of stocking can be disinguished by the purpose for which they are undertaken (Welcomme and Bartley, 1998). Stocking can be in compensation for the environmental disturbance caused by human activities like construction of dams across rivers for generation of elecricity, or for storage of water for irrigating agricultural crops. Stocking is of importance in maintaining aquatic populations by increased recruitment, to compensate for overfishing. In many instances it may become necessary to increase productivity of a water body at a higher level, by improving indigenous populations by stocking exotic sport fishes. Conservation of endangered species may also require stock enhancement by suitable means.

Chapter 3
Nature of Environmental Impacts

3.1 Conflicts with Other Uses

The nature and extent of environmental consequences of aquaculture depend largely on the location and type of farms, as well as the production technologies employed. The majority of present-day aquaculture production comes from inland and coastal pond farms and intertidal and foreshore open waters. While productive agricultural land is an ideal site for pond farms, such land is not easily available for aquaculture, and farms often have to be located on land that is described as wasteland or wetland.

As the siting of the farms has to be based primarily on access to surface or underground sources of water, the choice often falls on wetlands with a high water table, or on floodland areas. As will be discussed further in the following chapter, these areas are not truly wastelands as they have significant uses. Even when they are not used directly by man, they may have an important role in the maintenance of the ecosystem of the area and contribute to aquatic productivity and wildlife preservation.

Coastal wetlands are considered to be among the most productive natural systems known, and are sources of nutrients for organisms living in marshy areas as well as the water bodies into which they drain. The importance of coastal marshes as nursery and feeding grounds of the young stages of a number of commercial species of fish and shellfish has been well described in fisheries literature.

A good proportion of the existing land-based aquaculture farms is located on converted wetlands (see Fig. 3.1), and though quantified data are not available, it is reasonable to assume that these farms have brought about ecological changes and affected other uses of such areas. Conflicts with other uses and possible adverse consequences of converting wetlands into aquaculture farms are, therefore, major considerations to be addressed.

It has to be recognized that not all environmental consequences of aquaculture are negative. Indeed, many of them are highly beneficial to effective environmental management and the socio-economic wellbeing of human populations, when the land use is regenerative rather than

Figure 3.1   A mangrove swamp converted into a brackish water pond system (tambak) in Indonesia. (Photo by M. New)

merely exploitative. For example, dumping of domestic and industrial wastes near farm sites is more easily prevented and the multiplication of vectors of dangerous diseases such as mosquitoes and tsetse flies that thrive in marshy areas, at least reduced, if not eliminated. If domestic and farm wastes are used for fertilizing or feeding, not only would this constitute an inexpensive means of waste disposal, but it would be effective recycling to produce food and fodder.

Another major impact of land-based closed-type aquaculture systems is the abstraction, retention and drainage of water. Many farms depend on surface water from streams, rivers, lakes, estuaries and coastal areas, but some depend entirely on springs and underground sources. Competing use of surface water is largely restricted to irrigation systems and drinking water supplies. Though use of water in aquaculture is often considered as 'non-consumptive', there are significant losses due to seepage and evaporation, depending on soil properties and climatic conditions.

Aquaculture can often utilize water that is unsuitable for drinking and irrigation, such as saline waters occurring in semi-arid areas. In small-scale homestead-type fish farming, fish ponds often serve also as sources for crop irrigation, watering of livestock and for household uses. Integrated rice-field fish culture is another example of non-competitive use of water resources.

On the other hand, when sub-soil water has to be pumped out in large quantities, the water table of the area can, in the long run, become low, affecting adversely the underground water resources. When brackish or salt water is used to irrigate the farm, salt penetration of the soil in areas surrounding the farms can render them unsuitable for several types of vegetation, crops and trees. When farms are constructed along water-fronts near rivers, open estuaries and the sea, clearing of mangroves or other vegetation without leaving adequate protective barriers can result in soil erosion.

3.2   Sedimentation and Obstruction of Water Flows

In coastal culture of shellfish and cage farming, sedimentation and obstruction of coastal flow are important consequences to be con-sidered. The sediment may consist of fine particles of organic detritus, or coarse sand particles derived from the processes of erosion. Mollusc aquaculture in intertidal areas often faces severe problems of active sedimentation, which may result in the abandonment of beds or the transfer of beds towards the sea.

In traditional forms of bottom culture, horizontal beds of deposits are formed and they seldom affect the pattern of water flow. But in the rack culture of oysters (see Fig. 3.2), where bags of oysters are placed on racks arranged in lines hundreds of metres long and parallel to the tidal currents, sediments accumulate beneath and between culture racks. In the 'bouchot' type of mussel farming on poles driven into the seabed, as practised in France (Pillay, 1990), the lines of poles act as walls against the circulation of water when the mussels grow to adult size and block the spaces between the poles. There is heavy sedimentation and part of the sediments may be resuspended during succeeding tide flows and transported to the poles where they are redeposited. Eventual sub-mersion of the poles to as much as about one-half of their length has been observed (Ottman & Sornin, 1985). Studies made in Sweden show that the sedimentation rate in mussel farms can be three times that found in areas away from the farm (Dahlbäck & Gunnarsson, 1981).

In addition to the deposition of detritus that affects the pattern of water flows, the production of biodeposits by filter-feeding bivalves under culture augments the rates of sedimentation (see Section 7.2). The biodeposits utilize considerable quantities of oxygen in oxidizing the organic matter contained in them, and eventually create a reducing environment and the production of hydrogen sulphide (H_2S). In addi-tion, shear velocity and viscosity of mud are increased. Biodeposits increase the quantity of mud, and since organic mud is resistant to erosion, sedimentation rates are enhanced, resulting in the elevation of the seabed to the extent of 30–50 cm per year.

Cage culture also contributes substantially to the production of detritus and sediment deposits, which if not flushed out and dispersed by currents

Figure 3.2 Shellfish culture on racks that obstruct coastal flow and promote sedimentation.

or other means, can accumulate and cause pollution problems affecting not only the outside environment, but also fish production in the cages themselves. Organic loading can stimulate H_2S production and a reduction in the diversity of benthic fauna (Mattsson & Linden, 1983). The accumulation of fish faeces results in the increase of ammonium nitrogen and phosphate-phosphorus concentrations in the surrounding water column up to double and quadruple levels, respectively (Larsson, 1984).

3.3   Effluent Discharges

In response to environmental concerns relating to aquaculture practices, substantial attention has been devoted in recent years to the effects of discharges of effluents from certain types of aquaculture. In the majority of pond farms, discharge of water usually occurs only before harvesting of stock, which in many cases may be once a year or once in two or three years (see Fig. 3.3). The particulate matter and nutrients brought in with water supplies to the farm, together with the fertilizer, feed spill and waste products, interact with the pond soil and water to produce food organisms for the cultivated species. Excess detritus that accumulates on the pond bottom is removed after harvest to prepare the ponds for the next crop (see Fig. 3.4). Water quality requirements on the farm depend on the species cultured, but as far as possible, the quality of water is maintained at least at the same level as in natural sources.

Most of the nutrients in the inflows of water and those produced in the ponds are utilized for organic production, while some are absorbed by the soil. In well-managed pond farms, wastage of feed is minimized, but what is wasted is mineralized along with accumulated faecal matter, adding to the nutrient content of the water for organic production. Effluents from such farms are unlikely to be significantly different from the influents. For example, Boyd (1985) found that there was no accumulation of nitrogen and organic matter in sediments in channel catfish ponds, even though only 25–30% of the nitrogen, phosphorus and organic matter applied as feed was harvested as fish.

Figure 3.3 Coastal ponds being drained for drying the detritus on the pond bottom.

Figure 3.4 Excess detritus from fish ponds being sun-dried for use as agricultural fertilizer.

Figure 3.5 An intensive pond farm that has a high rate of waste production.

But, as will be discussed in Chapter 4, in pond, tank and raceway farms adopting intensive culture practices that include high rates of stocking, feeding, and water flow, the character of effluents can be different (see Figs 3.5 & 3.6).

Figure 3.6 Waste water being discharged from a raceway farm.

The wastes entering receiving waters from such farms consist of solid or soluble wastes. Solid wastes may be in suspended form or may accumulate on the sediment, and consist mainly of organic carbon and nitrogen compounds. The soluble wastes are generally derived from metabolic products of the cultured stock, or from the solid wastes through decomposition and leaching. The biochemical oxygen demand (BOD) of waste material, which is a measure of oxygen required by micro-organisms to decompose organic matter, is a valuable parameter for evaluating the polluting strength of wastes.

The effect of fish farm effluents on receiving waters varies considerably, depending on local conditions. This is evident from a comparison of the effects observed in national surveys in Finland (see Table 3.1) and in the UK (see Table 3.2).

3.4 Hypernutrification and Eutrophication

Hypernutrification and eutrophication are the two major processes that result from waste discharges from land- or water-based aquaculture farms. Any substantial and measurable increase in the concentration of dissolved nutrients is termed hypernutrification, and the consequent significant increase in phytoplankton growth and productivity (if growth is limited by nutrients) is termed eutrophication. In freshwater environments, where the term eutrophication has more commonly been used, dissolved inorganic phosphorus is considered the most important

Table 3.1   The effects of fish farm effluents observed in a survey of 200 fish farms.

Type of effect	Number of cases
Eutrophication	22
Increase in phosphorus content	15
Increase in sanitary indicator bacteria	11
Fall in dissolved oxygen content	9
Increased algal bloom	8
Increased 'sewage fungus' and settled solids	5
Increase in chlorophyll 'a'	4
Increased macrophytes	3
Increased turbidity of water	2
Smell	2
Bad taste in fish	2
Water not potable	2
Restricted use of potable water supply	1
Fish kills	1
Changes in benthic fauna	1
Polluted fish traps	1
Deteriorated fishery	1
Total	90

(From: Sumari, 1982)

Table 3.2   The effects of farm effluents observed in a survey of 141 fish farms in the UK.

Type of effect	Number of farms implicated
Suspended solids: deposition and discolouration	17
Algal blooms	3
'Sewage fungus'	3
Malachite green discharge	2
Lime	1
Smell	1
Total	27

(From: Solbe, 1982)

growth-limiting factor, and in the marine environment, dissolved inorganic nitrogen is the most important (Dugdale, 1967).

The quality of effluents from an aquaculture farm is significantly influenced by the annual production per unit volume of water and the retention time of water in the farm. Depth and temperature of the water, food supply and cleaning operations may also affect the quality of effluents.

The effects of discharge of effluents in the receiving waters are mainly an increase of suspended solids and nutrients and a fall in dissolved oxygen content. Aquaculture operations cause the release of metabolic waste products like faeces, pseudofaeces, excreta and uneaten food. Dissolved waste products like nitrogen and phosphorous remain in the receiving waters. The organic wastes reach the sediments. Most of the released solids, which are organic carbon, enrich the benthic ecosystem. Organic enrichment leads directly to an increased oxygen consumption rate, and when the oxygen demand is more than that which is available, the sediment becomes anoxic. This in turn brings about important changes in the biological and chemical processes in the sediment and the ecology of benthic organisms.

In areas of low turbulence and high organic input, the sediment/ water interface can also become anoxic. It has been observed that water under cage farms in turbulent locations can become depleted in oxygen for long periods (Gowen & Bradbury, 1987). Anaerobic processes develop when oxygen is depleted, and the sediment becomes reducing. This results in the production of ammonia, hydrogen sulphide and methane, which may be released into the water column. The rate of ammonia release under a freshwater cage farm was estimated by Enell and Löf (1983) to be between 2.6 and 3.3 times the rate of release from an undisturbed oxic sediment. The ammonia dissolves in the overlying water and adds to the total nitrogenous concentration. Most of the hydrogen sulphide (H_2S is slightly soluble in water) and all the methane produced are released in gaseous form. Out-gassing has been observed beneath many cage farms, leading to gill damage and increased mortality of fish.

Constant deposition of large amounts of waste can create azoic zones devoid of macrobenthic organisms beneath cage farms. An impoverished microfauna, dominated by opportunistic species characteristic of enriched sediments, may develop in the immediate vicinity of the farm. A transition zone with lower levels of enrichment and stimulated growth can be recognized beyond this, and further beyond a zone of normal conditions.

Hypernutrification generally leads to enhanced primary production and a standing crop of phytoplankton, although under certain circumstances this may not occur. It is likely that at higher levels of hypernutrification, other factors may limit primary production. The utilization of nutrients by phytoplankton may be affected by turbidity and availability of light, as well as currents and flushing time. Increased primary production may not result in an enhanced standing crop if the additional biomass is rapidly removed by factors such as grazing. Algal blooms, especially of toxic species produced by high levels of nutrients, can cause environmental hazards, including mortality of fish (see Fig. 3.7). Such algal blooms are likely to be localized and not of the catastrophic scale that has occurred in Norway and Scotland in recent years, which do not appear to have been caused by aquacultural wastes (see Section 7.7).

The biological effects of effluents from land-based pond farms have been studied in considerable detail in Denmark as a consequence of the environmental problems created by trout culture practices. Markmann

Figure 3.7   Algal bloom caused by hypernutrification near a pond farm.

(1982) lists the chemical and physical changes observed downstream of fish farms on Danish streams as follows:

(1)   more fine-grained and homogeneous sediment;
(2)   augmented concentrations of suspended organic solids serving as food for invertebrates, and of dissolved organic substances serving the same for bacteria and protozoa;
(3)   reduced concentrations of dissolved oxygen due to respiration of fish within the fish farm and to increased BOD and sediment respiration;
(4)   augmented concentrations of ammonia, nitrate and phosphate;
(5)   potentially toxic concentration of ammonia.

Increased levels of dissolved organic matter cause an increase in the number of micro-organisms, especially bacteria. In streams heavily loaded with fish farm wastes, the vegetation was often found to be dominated by filamentous algae. Where the vegetation includes macrophytes, they are often covered with diatoms and other single-celled algae. The species and growth of macrophytes depend on the grain size and homogeneity of the sediment.

In marine environments, changes in the natural population of phytoplankton and macro-algae have been recognized in the vicinity of farms where there is vigorous flushing. Besides stimulation of primary production, changes may occur in the species composition and abundance of phytoplankton and in enhanced macrophyte growth. Senescence and disintegration of phytoplankton blooms can cause areas of low dissolved oxygen.

The nature of fauna downstream of fish farms is obviously influenced by the dissolved oxygen concentration, the nature of substrates and availability of food as affected by the fish farm effluents. Populations of filter- and detritus-feeding invertebrates show better growth in increased concentrations of particulate organic matter and in sediment enriched with organic matter. The fauna in areas with low oxygen concentration downstream of farms are dominated by invertebrates that can regulate their oxygen uptake independently of concentrations in the surrounding water.

The change in species composition from upstream to downstream is expressed numerically as the saprobic degree or index, as proposed by Sladeck (1973; quoted by Markmann, 1982), to indicate the degree of pollution. Increases in the degree of saprobity downstream are indicative of the deterioration of water quality downstream and the elimination of 'clean-water species'.

Water-industry biologists in the UK use what is known as the Trend Biotic Index (TBI) to describe the state of benthic invertebrate communities in water bodies (Woodiwiss, 1964). The index varies from class X, which represents very good quality water supporting several species of stonefly (Plecoptera), to class I, representing sites suffering from gross organic pollution and able to support only a few very tolerant species, including those not relying on dissolved oxygen for respiration. A survey of fish farms in the UK, reported by Solbe (1982), showed that farms were generally sited on rivers with TBI values of IX to X, with very little change in classification downstream of fish farms.

Another approach to determining the effect of effluents in receiving water bodies is by comparing the so-called Chandler Score (Chandler, 1970), which assigns to each family, genus or species a score that reflects the sensitivity of the group to pollution, and fluctuates with its relative abundance. Large numbers of a very sensitive group would score high, while large numbers of a very tolerant group indicative of a stressed community yield low scores. The application of this approach in the UK survey of fish farms, however, showed that large farms producing more than 100 tons per year did not cause a reduction in the Chandler Score downstream, probably because they were located on large rivers.

Changes in fish stock and fisheries of the receiving water bodies is another means of assessing the impact of fish farm wastes. Changes in the water quality can be expected to change the composition of the fish resources, particularly of species that require higher quality water and oxygen concentrations. In surveys of the effects of fish farm effluents in Europe (Alabaster, 1982), there was no change in status of the fisheries in the majority of cases, probably because the impact of effluents was too slight and localized to alter the composition of the fish stock.

The University of Stirling *et al.* (1990) reports that there is evidence that fish farm effluents may stimulate fish production below fish farms in rivers and in some freshwater lochs. No adverse effects were found on downstream fisheries in effluent-receiving rivers in a number of European countries when the total flow was not less than 5 l s^{-1} for each tonne of annual fish production (Alabaster, 1982).

The impact on the environment is more direct in pen and cage culture and is pronounced when large numbers are concentrated in protected areas with insufficient water exchange. Unlike many other types of aquaculture, the environmental impact of cage farming has been the subject of scientific attention in recent years, especially in northern European countries where cage farming of salmonids has become a major industry. The presence of cages and pens in open waters does obstruct current velocity and increase sedimentation. According to Edwards and Edelsten (1976), the current speed within a cage may be half of that outside. Inone (1972) has reported that a net cage of $20 \times 20 \times 6$ m, with a mesh size of 50 mm, stocked with yellowtail at the rate of 1.6 kg m^{-3}, can reduce the current velocity inside the cage by up to 35% of that outside. A farm consisting of several such cages can have a very significant impact on the current pattern of the area.

The major ecological effect of cage farms is considered in the context of eutrophication of the water body, determined from the basic mass-balance of limiting nutrients such as phosphorus and nitrogen. There are differences of opinion on the importance of phosphorus, which has been established as a limiting nutrient in freshwater environments, versus nitrogen, which is probably the limiting nutrient in marine eutrophication. However, it is considered more meaningful to recognize the importance of both the elements, the forms of the nutrients, whether dissolved or particulate, together with the site and climatic season of discharge (Håkanson *et al.*, 1988).

The total discharge of nutrients from a fish cage is estimated to be generally in the order of 10–20 kg phosphorus (P) and 75–95 kg nitrogen (N) per year per ton of fish produced (Enell & Löf, 1983). This is based on data from north European countries and the type of food presently used for salmonids, and on the total quantity of fish produced rather than the number and size of fish at a given time. The feed conversion coefficients are calculated in the range of 1.6 to 2.2. Based on these figures, it is roughly estimated that the total discharge from a cage fish farm producing 50 tonnes per year would correspond to discharges from a sewage purification plant of a town of about 7000 inhabitants, assuming a 90% reduction of phosphorus in the plant (Håkanson *et al.*, 1988).

3.5 Chemical Residues

Besides the wastes derived from natural processes and wastage of feeds, the effluents from an aquaculture farm may contain remains of chemicals used to disinfect the farm, control pests and predators or treat diseases (Bjorklund *et al.*, 1990). Hormones may be used for induced breeding or sex reversal, and anaesthetics for facilitating the handling and transfer of fish. The nature and extent of the use of chemicals depend very much on the locality, the nature and intensity of culture operations, and the frequency of occurrence of diseases.

The concentrations discharged from the farms could be estimated only for malachite green for 24 farms (0.61 ± 0.70 mg l^{-1}), formalin for 17 farms (15.2 ± 48.1 mg l^{-1}) and hyamine for 4 farms (0.55 ± 0.97 mg l^{-1}). These values show wide variations but in relatively small amounts. From available information, the effects of discharges of the chemicals in the effluents are not considered to be significant in the receiving waters, as in most cases the discharge is only occasional and in low concentrations.

Anaesthetics, disinfectants and biocides used in farms may have lethal or sub-lethal effects on non-target organisms in the environment. There is the possibility of generating drug-resistant strains of pathogens by the use of antibiotics for treating diseases in farms and discharging the residues into the environment, though the resistance may be only short-lived. As the resistance to antibiotics can be transmitted from one bacterium to another, it has been suggested that there is a risk of transference of antibiotic resistance to normal bacteria in the human gut if antibiotic-resistant bacteria are ingested in numbers.

Even though the possibility of transmission of pathogenic infections from diseased farmed fish to wild fish has been suggested, there appears to be no evidence of it having occurred. On the other hand, there are several cases of infections transferred from wild stocks to farmed fish.

Biocides are commonly used in pond farms to control predators, pests and weeds, generally before the rearing begins. As no discharges are made from the ponds before the toxicity disappears, there is very little likelihood of the toxins reaching the outside environment. The environmental impact can be direct and detectable when chemicals are used to control pests and predators in mollusc farms and in cage and pen farming. Only localized effects have been observed so far, but there is a lack of information on their biological potency in micro quantities.

Besides these, the potential effect of chemicals derived from farm construction materials and antifouling compounds used in treating net cages have to be considered. Tributyltin (TBT) has been observed to cause reproductive failure in oysters and toxicity to other forms of marine life. It can also accumulate in the tissues of fish in cages.

3.6   Other Effects

The possible environmental effect of indiscriminate introduction of exotic species has been referred to in Chapter 1. The release of exotics as well as farmed fish into the environment can have adverse effects.

Besides predation or competition with local fauna, there are dangers of hybridization and reduction of genetic diversity. The inadvertent introduction of pathogens and diseases carried by the exotics is a particularly serious danger if adequate precautions are not taken.

Human health hazards include the transmission of water-borne diseases and the contamination of aquaculture products by human pathogens and toxic substances.

Though listed here as the last of the major environmental impacts of aquaculture, the diminishing of the scenic beauty of waterfront areas and the likely reduction of aquatic recreational facilities are of special importance because of their social implications and influence on public attitudes. The organizational strategy adopted for development can have major socio-cultural and economic impacts on local populations. The advent of large-scale farming in some countries has had the effect of phasing out small producers.

The impact of aquaculture on birds and aquatic mammals has not been adequately studied. However, there is enough evidence to show that disturbances caused by the location of farms near the feeding and breeding grounds of these animals, and the anti-predator measures taken by farmers, can affect their population size or established habitats.

Chapter 4
Extent of Environmental Impacts

It is often difficult to determine the impact of aquaculture on the environment in isolation, as the observed consequences are in many cases the cumulative effect of several factors that disturb its natural state. Available data seem to indicate that the pollutive effects of aquaculture are comparatively small and highly localized. In pond culture systems, the water quality of the outflows is not very different from the inflows, especially when retention time is long. In view of this, environmental control agencies such as the Environmental Protection Agency (EPA) of the USA do not appear to enforce any rigid federal quality standards for effluents from freshwater systems (Stickney, 1979). On the other hand, controlling agencies in the USA, Europe, Japan and many other countries regulate the siting and operation of marine farms to reconcile conflicts in the multiple use of land and water in coastal areas, and to prevent environmental degradation.

The destruction of mangrove swamps, which has created considerable concern in certain areas, is not all due to the conversion to aquaculture farms. These areas have also been used intensively for logging and are reclaimed for real estate and industrial development. There is no doubt that aquaculture has contributed to the changes, but its share of the impact is difficult to determine in quantitative terms. Conversion of mangroves into brackish water farms has been in practice for centuries. In the Philippines, over 206 500 ha, representing about 45% of the original mangroves, are reported to have been reclaimed for milkfish and shrimp farming by 1984. One-third to one-half of the 40 000–50 000 ha of shrimp farms in Ecuador are built on mangrove areas. The extensive brackish water pond farms, or tambaks, in Indonesia that cover over 225 000 ha were originally mangrove swamps.

4.1 Quantification of Effluent Discharges

Very few countries or regions appear to have any precise data on the quantity of effluent discharges or the total load of the major elements of nitrogen and phosphorus released into the waterways. Håkanson *et al.* (1988) have attempted to estimate the production of the load of organic

Table 4.1   Estimated load of organic matter (nitrogen and phosphorus) produced by marine fish farming in Nordic countries in 1986 (in tonnes).

	Norway	Sweden	Denmark	Finland
Gross production	50 000	3000	3000	7000
Organic matter, BOD	22 000	1300	1100	3100
Nitrogen, N	4 600	280	230	650
Phosphorus, P	540	35	30	76

(After: Håkanson *et al.*, 1988)

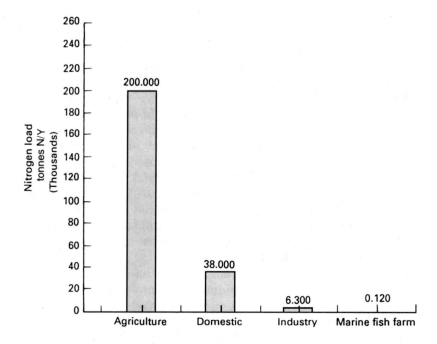

Figure 4.1   Nitrogen load derived from different sources in Danish seas in 1984 (From: Möller, 1987)

matter (nitrogen and phosphorus) in Nordic countries, based on an assumed feed conversion ratio of 1.65, as shown in Table 4.1.

The extent and significance of these loads are illustrated by a comparison of nitrogen loads from different sources (see Fig. 4.1). The total nitrogen exported from land to the Danish seas has been estimated to be about 150 000 tonnes per year, of which the load from sea farms is only about 0.2% (Håkanson *et al.*, 1988). The total nitrogen loss to the environment from agricultural production in Denmark is estimated to be around 460 000 tonnes per year.

These figures indicate the comparative magnitude of the problem in a country with a well-developed intensive-aquaculture industry, though the extent of the problem may not be the same in many other countries. Considering the north European countries as a whole, Leffertstra (1988) points out that the entire fish farming industry does not contribute significantly to nutrient loading of the North Sea, notwithstanding some of the observed local effects.

4.2　Assessment of Pollutive Effects

To assess the pollution potential of effluents from aquaculture farms, comparison with mass flows of pollutants from sewage works treating only domestic wastes has been suggested (Solbe, 1982). The mean discharge of BOD, ammoniacal nitrogen (NH_3–N), oxidized nitrogen (NO_3–N) and suspended solids from sewage works, calculated in terms of grams per person per day, can be used to express the mass flow of aquaculture farm effluents in 'population equivalents'.

For example, if the output per person of BOD is 2.48 g day^{-1}, an output of 1 g s^{-1} is derived from 34 800 people. An average fish farm discharging 0.3 g BOD s^{-1} can be equated with a sewage works treating the wastes of 10 590 people. Based on the mass outflow of 4.32 g person^{-1} day^{-1} of suspended solids from sewage works, the 'population equivalent' net mass outflow from fish farms can be expressed as that of 46 600 people.

However, objections have been raised to the use of population or person equivalent values as an index of the overall pollution load in fish farming (Rosenthal *et al.*, 1988) as fish farm waste water is not comparable to domestic sewage even though sewage is used as a fertilizer in certain types of fish farms. The individual components in treated sewage should nevertheless be comparable. The population equivalents of untreated sewage will be substantially different. For example, Bergheim *et al.* (1982) considered that pollution from 1 kg of trout is equal to that of 0.2–0.5 people. The loadings from a freshwater trout farm producing 200–250 tons of fish per year in Norway corresponds to untreated sewage from 1400–5000 people.

Having attempted to identify the main types and extent of environmental impacts associated with aquatic farming, the succeeding chapters will look more closely at aquaculture practices with reference to their individual and collective consequences on the environment.

Chapter 5
Siting and Design of Farms

As pointed out in Chapter 2, major environmental concerns relating to aquaculture development have focused on the occupation of land and water areas that should be preserved in their pristine state, and the effect of discharges from the farms. With a view to addressing these concerns and to avoid contravention of property rights and social needs, many governments require farmers and enterprises to obtain prior approval from appropriate agencies. While in many countries these approvals or permits relate only to ownership or to conflicts in land and water uses and community requirements, others also require approval of discharges and effluent limitations.

5.1 Restrictions on the Use of Potential Sites

It is not proposed to describe here the requirements enforced in all countries. However, a summary of some of the stricter regulations may be of interest in identifying site characteristics that have environmental implications.

The Environmental Protection Agency of the USA, in its proposed rules (Federal Register 39 (115), 1974a), requires that applications for approval of discharges from fish farm projects should include, among others:

(1) identification of the kind and quantity of pollutant(s) to be used in the aquaculture project;
(2) available information on:
    (a) the conversion efficiency of the pollutant to harvestable product,
    (b) the potential increased yield of the species being cultured,
    (c) any identifiable new product to be produced, including anticipated quantity of harvestable product;
(3) identification of the species of organisms to be cultured;
(4) identification of the water quality variables required for the growth and propagation of the cultured species including, but not limited to, dissolved oxygen, salinity, temperature and nutrients such as nitrogen, nitrates, nitrites, ammonia, total phosphorus and total organic carbon;

(5)   identification of possible health effects of the proposed aquaculture project, including:
  (a)   diseases or parasites associated with the crop which could affect aquatic life outside the designated project area and which could become established in the designated project area and/ or in the species under cultivation,
  (b)   the potential effect on human health,
  (c)   bioconcentrations in the crop including, but not limited to, radionuclides, heavy metals, and pathogenic organisms associated with the pollutant used,
  (d)   potential for escape of non-indigenous species from the designated project area;
(6)   identification of pollutants produced by the species under culture, especially those which may be channelled into waste effluent, such as ammonia, hydrogen sulphide, organic residues, phosphates and nitrates;
(7)   identification of the disposal method to be used, should there be a necessity for intentional destruction or massive natural death of the organisms under culture.

The State governments concerned are free to impose their own standards, although these may not be any more liberal than those of the EPA. Because of the difficulties and expense involved in carrying out the necessary tests, and the fact that most farms did not contribute significantly to the degradation of receiving waters, the requirements were limited to the determination of settleable solids. Apparently based on further studies which showed that most aquaculture effluents are of similar, or even higher, quality than that of the receiving water body, it now appears that no rigid federal quality standards for aquaculture effluents are imposed on freshwater systems (see Chapter 4). On the other hand, siting of marine farms is strictly regulated by a permit system, based on clearances by the different agencies concerned with environmental protection, navigation, commercial fisheries and the protection of sport fisheries and wildlife.

The possible conflicts with a larger number of other users, and more importantly the rapid increase in commercial interest and investments, have resulted in greater control on the establishment and running of marine aquaculture farms in the USA. The laws enforced by the States of Washington and Oregon are said to make it extremely difficult and time-consuming to obtain necessary permits. The State of Oregon even introduced a moratorium on river-hatchery development. Guidelines are now available to minimise the effects of netpen operation in habitats of special significance, and controlling sedimentation and water quality in Puget Sound (Washington State). These are expected to facilitate consideration of applications for starting cage and pen culture in the area. The water–land use conflicts, and navigation and aesthetic aspects of projects, have still to be considered at the county level (see Woodward, 1989).

The guidelines suggest an annual production limit of $110\,t\,km^{-2}$ and a minimum mean current velocity of $5\,cm\,s^{-1}$. The required depth is related directly to the annual fish production. An operation producing 10 tonnes annually should have 12 m depth; one producing 10–50 t, 15 m, and one producing more than 50 t, 20 m. If the current velocity can be increased to $25\,cm\,s^{-1}$, the depth can be reduced to 8 m, 12 m and 18 m, respectively, for the three production ranges mentioned.

Farming operations are not to be permitted on, or in the immediate vicinity of, habitats of eel grass, kelp beds, rocky reefs, clam populations, nursery areas, wildlife refuges, and bird and mammal habitats of special significance.

In Europe, it was Denmark that experienced the most serious results of inadequate control of siting, construction and operations of its trout farms. The deterioration of water quality and transmission of diseases made it necessary for the government to bring in legislative measures to regulate the establishment of new farms and the enlargement of existing farms. A farmer or entrepreneur has to obtain permission from the regional authorities (county councils).

The quantity of surface and/or ground water that can be used by the farm is determined in terms of volume per second and volume per year at the time of granting permission. The use of ground water may be limited to a certain period of the year. At dams constructed to divert water to the farm, surplus water should be available to enable the passage of wild fauna, including migratory fish. The farm design should facilitate control of pollution from routine operations as well as from accidents. The inlet and outlet channels of individual ponds (see Fig. 5.1) should be designed to prevent release of sediments into receiving waters,

Figure 5.1   A Danish trout farm, where individual outlets are now prohibited.

and direct outlets from individual ponds are prohibited. The maximum production limit, and in some cases the quantity of food to be used, may also be stipulated. The use of trash fish as feed, which was considered to be one of the major reasons for insanitary conditions and disease outbreaks, is no longer allowed.

The admissible limits of effluent constituents, including BOD, suspended solids and nutrients, are prescribed. Monitoring of discharges two to four times annually, together with spot tests by the authorities, is made compulsory. Dead fish are not allowed to be thrown into receiving waters. General guidelines for improving the operation of farms in order to avoid environmental pollution have been recommended by the Environmental Protection Agency. It is also compulsory for the farmer to keep records of fish stock, size of fish, amount of feed used daily, and medication of the fish.

Even in countries where there were no regulations that required environmental impact assessments for siting aquaculture farms, authorities have been obliged to enforce restrictions in response to political and public pressure. For example, in Scotland, demands for a moratorium on fish farm development were made on the grounds of scenic detriment, nearness of farms to ancient monuments and historic sites, adverse effects on tourism, and reduction in employment opportunities. Besides the destruction of scenic landscapes, aesthetic objections to fish farm development were given as the use of vividly coloured buoys, unsightly buildings and outdoor working areas associated with farms, and the occupation of remote and unspoilt areas. A special commission was therefore appointed to study the environmental effects of the industry and to develop siting and operational guidelines for its management. Recognizing the need for the support of both the industry and the general public for any management policy to be effective, a new consultation policy has been adopted, to take into account the views of all interested parties before an application for establishment of a farm is approved.

A set of guidelines has been formulated, instead of rigid regulations, to aid approval and management of enterprises (Crown Estate, 1987). For the purposes of approving farm sites, areas may be identified as sensitive or very sensitive. Sensitive areas are those where farming may affect other values and would require special area strategies for management. Farming will normally not be permitted in very sensitive areas, but if allowed, it will be subject to severe restrictions. The guidelines recommended for choosing a site are listed below.

1. Find sites appropriate to available technology or find technology appropriate to the sites.
2. Agree with other operators on desirable separation distances or come to joint arrangements which would permit closer siting.
3. Discuss land facilities with planning authority or find ways of servicing site without onshore development.
4. Limit the size of the site to the minimum necessary.

Figure 5.2   A cage farm situated in a deep Norwegian fjord, away from human habitation. (Photo by Ola Sveen)

5. Avoid important fishing grounds, scenic and wildlife areas, or take steps to ensure insignificant effects.
6. Select sites inconspicuous from scenic views, houses, etc., or adopt special design techniques. (see Woodward, 1989)

In India, where the Supreme Court is still considering restrictions on shrimp farming in coastal areas on the basis of public interest litigation, some of the southern states have ordered the closure of existing shrimp farms in coastal areas. In Nordic countries, coastal planning is restricted on the basis of carrying capacity and offshore cage farming has been introduced to reduce environmental damage. Zoning of land and water areas is another strategy followed by some south-east Asian countries; Malaysia identifies certain areas as aquaculture zones. Hong Kong, Japan, Korea and Singapore have designated marine fish culture zones (FAO/NACA, 1995).

In siting farms, the most attractive parts of scenic areas have to be avoided, insofar as it is practicable, and an adequate distance from such areas ensured (see Fig. 5.2). In determining landscape quality, the seascape from the land and the landscape from the sea have to be considered. The number of farms built in sensitive areas has to be comparatively small. Dispersal into smaller sites may reduce visual intrusion, conspicuous colours have to be avoided in farm installations, and equipment and the farms concealed as far as possible through landscaping with trees or hedges. Larger farms have to be separated at greater distances. It is also recommended that in cage farms, lay-out pattern of cages should be selected to suit the background, keeping superstructure to a minimum. Similarly, the lay-out of facilities on land has to be in character with the environment. Obviously, it is not easy to recommend or follow rigid rules relating to the maintenance of aesthetic values or scenic harmony.

5.2 Basic Data for Site Selection

A review of available information on the criteria employed and conditions stipulated for according permission for the establishment of aquaculture farms, even in a small number of countries, shows wide variations. While some of these criteria are obviously arbitrary or tentative, others are based on local environmental conditions and land-use patterns (ADB/NACA, 1998). Other than the nature and composition of expected emissions or discharges from the farm, the pattern of water exchange and the bottom dynamic conditions in the area have to be considered.

In the guidelines for site approvals, environmental sensitivity has been referred to as an important factor. Different areas may respond differently to the same nutrient dose. Besides the coastal volume in the case of marine sites and the water retention time, it is of importance to determine any areas of erosion, transportation or accumulation on the proposed site. If cages are installed in areas of erosion and transportation the accumulation of oxygen-consuming organic material will be minimized and spread over the surrounding waters and sediments. In contrast, in an area characterized by fine sediments and accumulation processes, the spread of organic materials will be restricted. If the bottom water exchange is small, oxygen deficiency will be higher, and this can lead to the formation of lethal H_2S.

Areas of erosion characterized by bottoms dominated by sand and harder materials are generally exposed to wind-generated waves and currents. Resuspension of fine materials occurs mainly in areas of erosion and suspension. During stormy conditions when the wave base is lowered, there is generally a large input of oxygen from oxygen-saturated surface water and an accelerated oxygen consumption by resuspended material. The loading of nutrients is also enhanced by resuspension. In such areas, the primary dose of nutrients may be used several times for bioproduction, as a result of internal loading of nutrients from the interstitial waters of the sediments.

Thus, the sensitivity factor of a given coast is a function of several variants, such as coastal volume, water retention time, mean depth and areas of accumulation, transportation and erosion, as a measure of the resuspension potential.

It is extremely difficult to determine directly the bottom dynamic conditions (erosion, transportation and accumulation) at a given site accurately, and so indirect methods have to be adopted. Håkanson *et al.*

(1988) suggested the use of the *in situ* cone penetrometer, which 'consists of three cones of different shape and weight, whose tips are adjusted at the sediment surface. The apparatus is lowered to the bottom by a cable and the cones are then allowed to penetrate the sediment for about five seconds. The cone areas are locked when the cable becomes taut on retrieval. After recovery, the penetration depths can be measured.'

It is a rough method of determination and not adequate in border areas between erosion, transportation and accumulation. The same penetration depths may be obtained in different bottom types, e.g. sand–gravel and shell. In such cases, sediment sampling and/or repeated measurements may be required.

It is easy to appreciate the importance of determining water retention time in an area in order to assess the possible effect of a projected aquaculture farm. But water exchange varies in time and space and is dependent on the topographical characteristics of the site. The retention time for surface water (above the thermo- and/or halocline) is generally much shorter than that for deep water. In some narrow and deep fjords, the retention time of deep water may be as long as 5–10 years, whereas in some open coasts, the deep water may be renewed by almost every storm.

Since oxygen deficiency develops primarily in bottom areas with heavy organic load, the determination of water retention time in deep areas is especially important. However, there appears to be no simple method of assessing it. Håkanson *et al.* (1988) briefly described the use of commercially available CTD-sonds (conductivity, temperature and depth), gelatin pendulums and dye tracers, gauges and morphometric formulae for this purpose.

Even though all the necessary basic data are not available, most countries involved in salmonid farming now have regulations to prevent crowding of farms, and insist on minimum depths and flows in water bodies that receive the waste discharges. The recommended distances between farms vary according to hydrological conditions and size and production capacity of the farms. For example, the separation distances in Table 5.1 are suggested in Scotland.

Even though a distance of 3 km between salmon culture operations has been proposed for the west coast of Canada, the minimum distance enforced in British Columbia is reported to be only 0.5 nautical miles (800 m). In Norway, a 1 km separation is suggested by veterinarians to minimize the potential for transfer of diseases.

Some countries have declared the maximum admissible production in marine salmonid farms that can be allowed in the environment, obviously based on waste discharges. For example, on the east coast of Canada, production per farm is limited to 75–100 t of marketable fish per year. In Finland, there is generally a limit of 30–50 t per year. In some prefectures in Japan, as in Mie, there is a limit on production of 40 t ha^{-1} per year. The maximum size of cages for growing salmon and other fish in Norway is provisionally set at 8000 m^3 in the sea and 5000 m^3 in freshwater, although the maximum production is not specified.

Table 5.1 Suggested separation distances in miles (= 1.609 km) for sea farms in Scotland.

	Distance to	
	Salmon farms	Shellfish farms
Salmon farms	5	2
Shellfish farms	2	1
Viewpoints	1	0.5
Hotels/tourist centres	1	0.5
Houses	0.5	0.25
Wildlife colonies	0.5	0.25
Anchorages/approaches	0.25	0.25
Fishing grounds	0.25	0.25

(From: Crown Estate, 1987)

Some countries stipulate the areas where farming can be undertaken, based on site characteristics including depth and exchange rate of water. Cages are generally required to be installed at least 4 m above the seabed in marine systems. All these conditions are intended to ensure adequate dilution and distribution of effluents.

5.3 Siting Farms on Marshes and Mangroves

As mentioned earlier, many land-based coastal aquaculture farms are located on salt marshes and mangroves. Large areas of coastal waters are covered over by dense beds of several species of seagrasses, which perform important biological and physical functions in the marine environment. They stabilize the substrate, produce sediments and serve as habitats, nurseries and primary food source for many commercial and forage organisms. Coastal marshes, which abound in temperate latitudes, are highly productive, both where higher plants grow and on the mud and sand flats where only algae grow. Most of the vegetation is turned into detritus, which is a source of food for organisms within the marshes as well as in the estuaries and bays into which they drain. The marshes help in controlling erosion, and also serve as buffer areas to prevent flood damage. They form the nursery grounds of a number of species of commercially important finfish, shrimps and crabs. Oysters, mussels and clams form beds in tidal areas. The marshes are valuable as flyways, nesting and over-wintering grounds for many species of birds, as well as in harbouring several species of mammals and amphibians which need to be conserved.

The rate of reclamation of marshes, and particularly mangrove swamps, has accelerated in some parts of the tropics due to the rapid expansion of pond farming of shrimps for export. As mentioned in Chapter 4, about 50% of the mangrove forests in the Philippines have been developed into brackish water fish ponds (Saclauso, 1989). The area converted in

Thailand is estimated to be about 27% and in Ecuador about 13–14%. Such large-scale conversions have aroused considerable environmental concern among the public and development agencies. Though very few critical field studies have been carried out on the effect of these conversions, several seminars and discussions have been held, which have served to bring together existing information on tropical mangroves.

Mangroves grow throughout the tropics and, to a lesser extent, in subtropical regions on sheltered shores bearing soft intertidal sediments. North of the equator, mangroves extend to approximately latitude 30° on the east coast of North America (Louisiana), to 30° on the Pacific coast (northwest Mexico), to 32° in the Bermuda islands, to 27° on the Red Sea coast (Gulf of Suez) and 31° in Japan (Kyushu) (Macintosh, 1982). In the southern hemisphere, mangroves reach as far as latitude 32–33° south on the east coast of Africa. Their extreme southerly limit, according to Macintosh (1982), is Corner Inlet (38°–45° S) on the Victoria coast of Australia.

5.3.1 *Multiple uses of mangroves*

Compared to other wetland sites, tropical mangrove swamps have greater direct use to man, even though in some areas they also harbour the carriers of fatal human diseases such as virulent forms of malaria and schistosomiasis. Mangrove ecosystems have been described as constituting a reservoir, refuge, feeding ground and nursery for many useful and unusual plants and animals. Probably the most important feature of the system, as in coastal marshes, is the export of large amounts of detritus by tidal currents to adjacent coastal waters, which form a major source of nutrients for maintaining productivity. For example, a stand of *Rhizophora apiculata* in southern Thailand was estimated to produce 7 t of leaves and 20 t ha^{-1} of wood per year. Some of this enters the foodweb as detritus. Phytoplankton production in coastal waters here is around 5000 mg C (carbon) m^{-2} d^{-1}. Heald (1971) estimated that 40% of the detritus in a south Florida estuary was exported from a mangrove swamp.

The intertidal areas of mangroves are used as breeding–nursery grounds by several commercially important species, especially shrimps. Consequently, these species are described as mangrove-dependent and, based on correlations of commercial fisheries with occurrence of mangroves in certain areas, a slogan has been raised: 'no mangroves, no prawns' (Macnae, 1974). The correlation is actually not so clear, as in other areas there are important shrimp and prawn fisheries even though most of the mangroves have been cleared. Silas (1987) points out that not all shrimps use the mangrove environment equally, and that some frequent mangroves as juveniles, and some do not enter mangroves at all. It is still an open question whether those that occur in mangroves can survive in shallow coastal areas after the mangroves are cleared.

Though considerable published information is available on the productivity of mangroves, knowledge on the quantitative and functional

Table 5.2 Scales of impact commonly encountered with destructive uses of mangrove ecosystems.

Activity	Scale of impact (ha)
Clear felling	10 000–500 000
Diversion of fresh water	1 000–500 000
Conversion to agriculture	100–100 000
Conversion to aquaculture	100– 10 000
Conversion to salt ponds	100– 1 000
Conversion to urban development	100– 1 000
Construction of harbours and channels	100– 1 000
Mining/mineral extraction	10– 100
Liquid waste disposal	1– 10
Solid waste/garbage disposal	1– 10
Oil spillage and other chemicals	1– 10
Exploitive traditional uses	1

(From: Saenger *et al.*, 1983)

interactions in the system is extremely limited. In addition, it has been shown that not all mangroves and marshes are equally productive (Walsh, 1967; Nixon, 1980). Irrespective of this, it is generally accepted that mangroves constitute unique ecosystems that need to be conserved.

As mentioned earlier, there are several uses for the mangrove forests, and the International Union for Conservation of Nature and Natural Resources (IUCN), in its global review of the status of mangrove ecosystems (Saenger *et al.*, 1983), has identified 22 major uses. Logging is probably the most important economically. Because of this, many mangrove areas are managed as reserve forests, so that the resources can be regulated to ensure regeneration of the mangroves. However, destructive uses have become common for various reasons, and this is the major cause of concern. Even in the pattern of forest exploitation, clear felling for woodchips has become a major threat, as can be seen from the impact estimates of destructive uses listed in Table 5.2.

Clearly the impacts of any of these activities on a particular location are highly variable. Nor (1984) summarizes the main guidelines in the rational use of mangrove ecosystems as follows:

  '– do not alter the time and quality of freshwater runoff entering mangrove communities;
  – do not alter tidal inundation frequencies or surface circulation patterns;
  – do not alter physical structure, chemical properties and biological activities of the sediment substrate.'

It is necessary to consider the impacts of activities not only in the mangrove itself, but also upstream, inland and along the coast, which may lead to negative impacts.

Table 5.2 would seem to indicate that the scale of impact of conversion of mangroves for aquaculture is much less than for some other uses. Nevertheless, the cumulative impact of several uses can be much greater and more serious. It has also to be remembered that large-scale destruction of mangroves for aquaculture has so far occurred only in a small number of countries, where the extent of future conversions is now regulated. In the absence of adequate scientific data, arbitrary or tentative limits have been imposed.

There is a general recognition among development specialists that the possible adverse effects of each project should be individually assessed and the mangrove ecosystem should be managed on a sustainable yield basis, which considers social and ecological needs as well as economic ones. The lack of quantitative information on the economic value of the ecosystem when used on a sustainable basis, and the paucity of tested strategies for multiple-use development of marsh and mangrove areas, are major constraints to rational management.

Experiments in replanting for the rehabilitation of mangrove raise the possibilities of combining aquaculture with the regeneration of mangrove ecology (Fitzgerald, 1997; Phillips and Barg, 1999).

5.3.2 Mangroves as sites for farms

From the point of view of suitability for siting aquaculture farms, mangrove swamps would rank very low in the order of preference. Besides physical isolation, lack of communications and seasonally unsuitable salinity conditions, the major drawback is the occurrence of acid sulphate soils in most mangrove swamps. This is caused by the formation of pyrite, which is fixed and accumulated by reduction of sulphate from salt water. It involves bacterial reduction of sulphate to sulphide, partial oxidation of sulphide to elemental sulphur, and interaction between ferrous or ferric iron with sulphide and elemental sulphur.

The occurrence of sulphate, iron, a high content of metabolizable organic matter, sulphate-reducing bacteria (*Desulfovibrio desulfuricans* and *Desulfo maculatum*) and an aerobic environment, alternating with limited aeration in the mangrove environment, are factors that contribute to the production of sulphate soils. Such soils are more common in the zones between mean high water and mean low water levels, with limited periodic aeration due to tidal fluctuation. They are less prevalent in the better-drained parts of swamps, which are aerobic most of the time.

The reclamation of mangrove swamps into pond farms with drainage (see Fig. 5.3) results in exposure and oxidation of pyrites, and causes acidic conditions. Ferrous iron (Fe_2) is released during atmospheric oxidation of pyrites under an optimum moisture content of 30–40%. At low pH, ferrous iron is oxidized to ferric iron (Fe_3) by oxidizing bacteria. It can remain in solution in appreciable amounts only at a pH between 3 and 3.5, and is a more effective oxidant for pyrite and elemental

Figure 5.3 Mangrove swamp being reclaimed for shrimp farming. (Photo by Ms M. Loyche—courtesy K.C. Chong)

sulphur than free oxygen. At higher pH, all ferric iron is hydrolyzed and precipitated as ferric hydroxide. Basic ferric sulphate is also formed during pyrite oxidation. Elemental sulphur is oxidized to sulphuric acid by bacteria.

The most harmful effect of pyrite oxidation is caused by the excessive amount of sulphuric acid produced, which if not neutralized by exchangeable bases, creates strongly acid conditions. It is not only the existence of sulphate soils at the time of site selection that has to be taken into account, but also the potential for acid conditions to develop as a result of drainage when pond farms are constructed. Tidal brackish water vegetation with dense rooting systems is usually related to the accumulation of pyrite. Association of the red mangrove (*Rhizophora*), *Nipa* and *Maleuca* stands is a fair indication of potential acid condition of the soil. Besides acid conditions in the farm itself, drainage from dried embankments of the farm is likely to be acidic and may give rise to acidic conditions in the receiving waters in the neighbourhood of the outfall.

In spite of these unfavourable conditions, the only available sites for the development of coastal farms may be mangrove swamps in certain regions. Under such circumstances, the option would be to select areas of the mangrove swamp that would least affect the ecological value and resource use of the system, and allow the regenerative potential to continue its natural functions and sustainable use. This implies multiple uses, which involve the allocation of areas according to specific criteria and resource capabilities.

The zonation in mangrove swamps can be recognized on the basis of tidal inundation or the nature of vegetation. Macnae (1968) distinguishes the following six zones in his account of Indo-West Pacific mangroves:

- landward fringe,
- zone of *Ceriops* thickets,
- zone of *Bruguiera* forests,
- zone of *Rhizophora* forests,
- seaward *Avicennia* zone,
- *Sonneratia* zone.

Terchunian *et al.* (1986) recognized two zones in the mangrove forests of Guayaquil (Ecuador), namely the *Rhizophora* zone bordering the rivers and behind it a zone of mixed mangroves (usually *Avicennia* and *Laguncularia*). Zones of salt deposits (salinas) and mud deposits occur in many areas.

The more classical pattern of zonation based on tidal inundation is as follows:

- inundated by all high tides,
- inundated by medium high tides,
- inundated by normal high tides,
- inundated by spring tides only,
- inundated by equinoctial or other exceptional tides only.

These zone types may not be found in all mangrove areas as they are modified by climatic conditions, coastal morphology and freshwater outflow. However, in all swamps the fringe areas that are most frequently inundated by normal high tides, where sediment accretion occurs almost continuously, are the most favourable zones for primary growth of mangroves (see Fig. 5.4).

The first crop or primary growth of the red mangrove (*Rhizophora*) can only root in soft mud deposits that are dry for several hours a day. Once the mud banks have risen above low water they accumulate sedimentary and organic detritus around the prop roots at an accelerated rate. The mangroves grow tall and flourish, utilizing nutrients from the soil and water. Too high or too low water may hamper their growth. If not felled, these tall trees die naturally, and their place is taken over by a secondary growth of the same mangroves, but this does not develop beyond the stage of shrubs. Taller mangroves start growing in the new accretions in front of the secondary growth. The secondary growth appears to propagate not through seeds, but by root stocks or other vegetative means. The reason for the secondary growth remaining shrubby is obviously the loss of adequate quantities of fresh nutrients through sedimentary accretion and tidal flows. By the time the primary growth

Figure 5.4   Tall mangroves growing on the border of an estuary.

of red mangroves mature, the soil below becomes filled with minute rootlets, with the result that several centimetres of the upper layer look very much like thick peat. Back swamps usually have such soils, and are difficult to use for building pond dikes. Only in rare cases will the mangrove area be above high-water level, and in such areas the secondary growth dies and other forms of vegetation, such as ferns, take over.

Even though the area of primary growth is the most fertile part of a mangrove swamp, this is not suitable for farm construction because of the high cost of clearing and the need for conserving it for protection from erosion, tidal floods and natural disasters. The back swamps (see Fig. 5.5) with secondary growth of mangroves which contribute relatively little to the productivity of the swamps are the areas that should be investigated for siting farms. The cost of clearing the shrubby growths will be less, although excavation and removal of the peaty surface soil will, to some extent, reduce the savings. Drainage from the swamp will be improved by the tidal canals built for the farm.

In the absence of appropriate quantitative data on the extent of swamp areas that can be converted without detriment to the ecological processes and other uses, limits based on empirical calculations may have to be made. If the mangroves are rationally managed, facilitating adequate regeneration, it would appear that large areas can be made available for aquaculture, even in countries where large-scale destruction has taken place.

Figure 5.5 Secondary growth of mangroves in back swamps. (Photo by H.R. Rabanal)

From a preliminary assessment of the effects of constructing shrimp farms in the back swamp zone in Ecuador dominated by *Avicennia* and *Laguncularia*, Lahman *et al.* (1987) suggest that it has not only affected the habitat diversity, but also the occasional flushing of fine particulate and dissolved matter into the nearshore environment from this less frequently inundated zone during storms and neap tides. This could reduce the production of heterotrophic micro-organisms that form a food resource for crustaceans.

In order to minimize conversion of swamps for expansion of pond farming, the introduction of intensive systems of farming has been suggested (Kapetsky, 1987). This would involve not only major changes in the present farming systems, but the production and use of additional inputs and farming skills. More important for the environment is the increased loads of wastes in the discharges from the farms. This can, however, be controlled by appropriate waste treatment in the farms before discharge.

5.4   Selection of Water Bodies and Stocks to be Enhanced

Though the underlying principle of all types of stock enhancement is the maximization of levels of production, the categorization of types of enhancement or the development of culture-based fisheries is complicated by the variety of circumstances under which it is employed in fisheries. Enhancement of stocks may be undertaken in marine, coastal and inland waters. The techniques adopted and the species involved may depend upon the conditions that are created by natural causes or by human intervention.

Some of the earliest stock enhancements were undertaken to mitigate the declining populations of anadromous salmon and sturgeon that were affected by the damming of rivers across their route of migration, reduction of recruitment caused by over-fishing, and environmental perturbations such as water pollution and habitat, destruction, especially of spawning grounds. To compensate for reduction in recruitment many countries attempted to stock the system with young fish raised in hatcheries. They are released in the natural habitat, where they feed and mature before migrating to their home waters for spawning. Enhancement of Pacific salmon (*Oncorynchus* spp.) is an example of successful ranching that has contributed to salmon fisheries in coastal and riverine areas of the Pacific Northwest (Isaksson, 1988). Hatchery development and improvement of spawning beds/channels have been integral parts of the technology used. Another impressive achievement is the rehabilitation of the Caspian sturgeon (beluga) (*Huso huso*) and (*Acipenser* spp.) that had been adversely affected by the damming of the Volga and other rivers of the Caspian basin. Large-scale enhancement of Caspian sturgeon was also undertaken in Iran in order to compensate for diminishing stocks due to loss of spawning grounds in rivers entering the southern Caspian Sea. It is reported that all the present-day fishery is based on stocked fish.

Stock enhancement became a major government policy in many East European countries with centrally planned economies, as it was efficient in terms of worker productivity as well as in employment generation for the community as a whole. Large reservoirs formed by hydro-electric power stations in the former Soviet Union are managed as the locations of major sturgeon fisheries not only by stock enhancement but also by the transplantation of food organisms (Jhingran & Natarajan, 1979). In order to enhance the fishery potential of peninsular India and newly formed reservoirs for hydro-electric power generation in other parts of the country, quick-growing species of Indian major carp (*Catla catla*), rohu (*Labeo rohita*) and mrigal (*Cirrhina mrigala*) were stocked starting in 1960. Many of these reservoirs developed breeding populations of Indian major carp and some of tilapia. By auctioning fishing rights to private parties, many of the reservoir management authorities are able to generate revenues to compensate for the investments required to develop fishery resources and manage them.

The Chinese system of management of lakes and reservoirs is mainly an extension of fish farming on a large scale. This involves habitat improvement and elimination of predators. Natural spawning is promoted by provision of appropriate breeding facilities for indigenous or stocked populations. Stocking density is adopted on the basis of size of the water body and its production characteristics. Fertilization by organic or chemical fertilizers may be undertaken when higher stocking rates are adopted. Small irrigation reservoirs built in some countries in

Africa by the colonial powers are also maintained by adopting fish farming techniques of indigenous species that have shown amenability to aquaculture.

Stocking has been adopted for creating or improving recreational fisheries. Many of the cold streams in tropical areas have been stocked with rainbow trout by angling associations in Asia and Africa. After repeated stocking with imported eggs and juveniles, many of them have established successful hatcheries. Where they have been managed properly, trout culture has become a commercial activity. Some industrially developed countries have undertaken marine stocking programmes such as for striped bass (*Morone saxatilus*) and red drum (*Scianops ocellatus*). It is believed that stocking has contributed significantly in many areas, although the results are not always amenable to cost–benefit analysis.

Recent Japanese programmes of marine stocking of several species of finfish, shellfish and crustaceans were occasioned by the ratification of the UN *Convention on the Law of the Sea* that granted a 200-mile exclusive economic zone (EEZ). This has resulted in restrictions on Japan's foreign fishing rights, and it has launched a policy of developing fishery potential over its extensive continental shelf area. Stocking is claimed to be a major source (90%) of the chum salmon fishery, 50% of prawn (*Penaeus japonicus*) catches, yields of red sea bream, all of the scallops landed and a good proportion of red sea bream (*Sparus major*) and flounder (*Paralichthys olivaceus*).

This overview of the various types of stock enhancement undertaken in different parts of the world, and the circumstances in which each type of enhancement is employed to maximise fishery resources may provide the basis for selection according to conditions obtaining in various localities. It can seldom be employed without the cooperation of the local community, as the cost–benefit ratio and the ownership rights of enhanced stocks are not clear enough to be bestowed on individuals or even cooperatives without the state's partnership. Establishment of hatcheries or collection of stocking material and the management of enhanced stocks, including habitat improvement, will need policy decisions at a government level and legislative sanctions. Ungson *et al.* (1993) provide details of the organization and funding of red sea bream ranching in Kagoshima Prefecture, Japan. They concluded that the project was an economic success, showing a high rate (~15%) at a market price of US $35–115 kg^{-1}.

5.5   Farm Design

In site selection and farm design, particularly in the case of land-based farms, the general tendency is to pay special attention to the quantity and quality of water supplies to the farms, and very little to the discharge and dispersion of effluents. As indicated in Section 4.2, the main constituents of the effluents that are of importance from the viewpoint of environmental impact are suspended solids and dissolved ammoniacal nitrogen (NH_3–N), nitrate nitrogen (NO_3–N) and inorganic phosphorus. The dissolved oxygen concentration of the effluents depends on the retention time of water in the farm, and differs considerably between stagnant ponds and flow-through systems such as raceways and tank systems. The stock size in the farm also influences the dissolved oxygen concentration. Alabaster (1982) has reported an average change in concentration of 1.6 mg l^{-1} between water inlet and outlet, for an average flow of water of 1.72 l s^{-1} t^{-1} annual production of salmonid fish in European farms. These values will be different if aerators are used for forced aeration.

5.5.1　Water exchange

It has been pointed out earlier (Section 3.4) that the impact of effluents depends on their quality and the rate of their dilution in the receiving waters. The quality of the wastes can be improved by appropriate changes in culture technologies (which will be discussed in Chapter 6) and by the addition of waste treatment facilities in the farm lay-out. But the dilution factor governed by water exchange in the receiving water has to be determined as a part of site selection. Water retention or exchange is greatly influenced by a number of hydrological and seasonal factors. If the recipient of effluents is a stream or river, variations in water level and flow rates can be highly seasonal depending on precipitation and other sources of water, such as springs and melting snow.

In the case of coastal waters, the tides, freshwater discharges, winds, thermal effects and coastal currents are important factors that influence water exchange. Fluctuations in the boundaries of thermoclines that separate warmer and lighter coastal water from colder and denser water, and the boundaries of haloclines that separate saline and denser water from lighter freshwater, may also affect surface and deep water retention time, particularly in relatively small and shallow coasts. These fluctuations greatly influence the spread of fish farm effluents. As it is seldom that a farming enterprise can bring about major improvements in the recipient waters to facilitate adequate dilution of effluents, it is essential that all relevant factors are investigated before selecting the farm site in order to assess flow conditions at different times of the year.

5.5.2　Waste treatment facilities

The concentration of waste elements can vary between hatcheries and production farms, while the constituents remain essentially the same. The design of waste treatment facilities can also be similar, even if the size often has to be different because of the volume of effluents to be treated. The typical waste water treatment methods are sedimentation,

Table 5.3 The quality of effluent from flow-through fish farms, river water and biologically treated domestic waste water (mg l^{-1}).

	River water	Fish farm effluent	Domestic waste water
BOD	1.0–5.0	3.0–20	300
Total N	1.0–2.0	0.5–4.0	75
Ammonia-N	–	0.2–0.5	60
Total P	0.02–0.10	0.05–0.15	20
Suspended solids	–	5–50	500

(From: Warrer-Hansen, 1982)

biological filtration, chemical treatment for precipitation, mechanical filtration, and others such as ion-exchange (Warrer-Hansen, 1982). Large quantities of water are utilized in aquaculture farms and most is discharged after varying periods of retention in the farm. For example, salmonid farms require a water flow of about 12–20 l s^{-1} t^{-1} of production. This is probably more per unit of production than for most manufacturing processes. The cost of advanced treatment of such large quantities of water is prohibitive for both large and small-scale aquaculture farms.

Experience so far in water re-use systems suggests that sedimentation, followed by biological nutrient removal, is likely to be the most feasible means of fish-farm waste treatment (see Liao & Mayo, 1974; Muir, 1982). Willoughby *et al.* (1972) and Warrer-Hansen (1989) consider sedimentation to be adequate and the most economic form of waste treatment for trout hatcheries and production farms.

To rationalize the selection of treatment methods, it is useful to compare the characteristics of aquatic farm wastes with domestic waste water and the water of recipient water bodies. Table 5.3 attempts such a comparison of fish-farm effluents with water of an unpolluted river and biologically treated domestic waste water in Denmark (Warrer-Hansen, 1982). From the table it is clear that effluents from flow-through fish farms can be considered a relatively clean waste water.

Reduction of BOD

Liao (1970b) reported that normal salmonid hatchery effluent contained 5 mg l^{-1} BOD, which is usually lower than the BOD in the final effluent of an efficiently operating secondary municipal waste treatment facility. Willoughby *et al.* (1972) compared the BOD of a municipal sewage system in the USA with that of a trout hatchery. A municipal sewage system carries up to 250 ppm BOD to its treatment facility. Generally, secondary treatment standards require 90% reduction in both settleable solids and BOD before discharging the effluent into public waters. So after secondary treatment there can be a BOD level of 22.5 ppm left for discharge. Compared to this, a 90% reduction of settleable solids, which removes 85% of the BOD through any form of treatment of the hatchery effluent, will leave only 0.5 ppm (90% of 5) of BOD. They estimated that the 90% BOD level of 5 ppm can be achieved by 12 minutes of settling in a settling basin. In view of this and the costs of advanced water treatment, sedimentation is the recommended treatment for hatchery and fish farm wastes. It is a widely used treatment method for wastes with low concentrations of pollutants in large volumes of water.

Settling of suspended solids

The main sources of settleable solids are the faecal matter of fish or shellfish cultured in the farm and the wastage of feed provided. If the incoming water contains detrital matter, this may add to the quantum of settleable solids. Increase in the concentration of waste material at times of cleaning the holding facilities, or partial or complete harvesting, has also to be taken into account. For example, Liao (1970b) reported that the BOD of salmonid hatchery effluent can rise from the normal 5 mg l^{-1} to as much as 49 mg l^{-1}. Boyd (1978) reported up to 100 ml l^{-1} of settleable matter in effluents discharged from catfish ponds after seining. The settling velocity depends to some extent on particle size. Resuspension in farms may occur as a result of the activity of the cultured species, for example feeding, as well as the agitation of water by routine culture operations such as sample fishing, aeration etc. Resuspension results in a decrease of particle size, with fine particles settling more slowly. While resuspension generally occurs in earthen ponds, it seldom happens in specially designed tank farms. Thus, suspended solids in tanks settle more rapidly.

The nature of effluents naturally vary with the farm design, species cultured, culture practices and quantity of water used. If the concentration of suspended solids and BOD exceeds permissible limits, it is advisable to build the necessary settling facilities in the form of a basin, tank or pond. A typical settling tank or pond would have an inlet, and settling and outlet areas. The sludge accumulates below the settling area (see Fig. 5.6). The design of the settling pond has to be based on the characteristics of the wastes, including the structure and settling

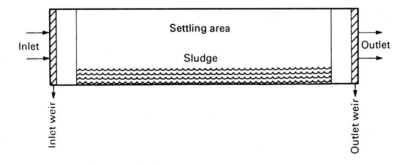

Figure 5.6 Lay-out of a settling pond.

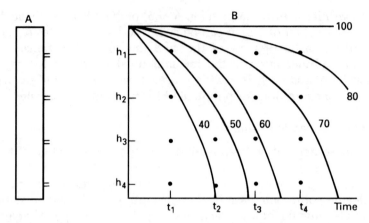

Figure 5.7 Diagram of a settling column (A) with 4 sample valves at heights h_1 to h_4 and an example of a settling rate/time relationship (B) for times t_1 to t_4. Points in B indicate observations, and the lines with numbers the interpolated percentage removal of suspended solids. (From: Warrer-Hansen, 1982)

properties of the solids, as these vary with the farming practices and the quantity of water utilized.

The settling characteristics of the solids can be determined by batch sedimentation tests in specially constructed settling columns, with sampling valves at different heights (Warrer-Hansen, 1982). The column is filled with a sample of the waste water, and after an initial sample is taken for analysis of suspended solids, further samples are taken at different times, for example, after 5, 10, 15 and 30 minutes. The analyses of these samples for suspended solids would provide the data for developing a series of curves relating settling rate to time, as shown in Fig. 5.7. This information can be converted to a proportion of solids with a settling velocity equal to a series of values (Warrer-Hansen, 1982).

The overflow rate ($m^3\,m^{-2}\,h^{-1}$) and retention time for various percentages of removal can also be estimated. It is considered necessary to apply a safety factor of 1.5 to the estimated overflow rate to compensate for the effects of turbulence. For example, if the overflow rate is $3.6\,m^3\,m^{-2}\,h^{-1}$, estimated on the basis of a settling velocity of $3.6\,m\,h^{-1}$ of the particular fraction of the particles to be removed, it should be adjusted to $2.4\,m^3\,m^{-2}\,h^{-1}$. The required area of the settling tank–pond is derived by dividing the volume of water utilization by the adjusted overflow rate. For example, if the water utilization is $1200\,m^3\,h^{-1}$, the required settling area will be:

$$\frac{1200\ \mathrm{m^3\ h^{-1}}}{2.4\ \mathrm{m^3\ m^{-2}\ h^{-1}}} = 500\ \mathrm{m^2}$$

Though this determination does not take into account the depth of the tank–pond nor, therefore, the retention time, it is necessary to ensure

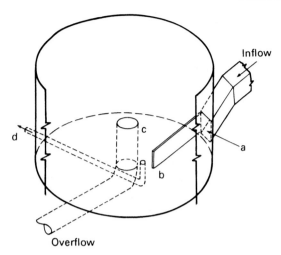

Inflow

Overflow

Figure 5.8 A swirl-concentrator showing waste water inlet (a), flow deflector (b), overflow of clarified effluent (c), and outlet for separated solids (d). (From: Warrer-Hansen, 1982)

that the water velocity in the settling zone is not too high, as otherwise resuspension of the settled solids may occur. Based on experience, Warrer-Hansen (1982) suggests that the velocity should not exceed 2–4 cm s^{-1}, and the depth and retention time have to be adjusted to obtain this velocity.

Based on Danish experience, Mortensen (1977) concludes that a detention period of 20 minutes in a settling tank is sufficient in most cases to obtain satisfactory settling of suspended solids. However, Henderson and Bromage (1988) found that at concentration levels of suspended solids <10 mg l^{-1}, efficiency of sedimentation in basins is greatly reduced and it is difficult to attain effluent concentrations of <6 mg l^{-1}. When solid levels are low, resuspension of particles occurs within the settling basin, and this accounts for a large percentage of solids in the outflow. Henderson and Bromage (1988) report that the settlement can be optimized for a given inlet solids concentration by maintaining a low mean fluid velocity of <4 m min^{-1} to minimize turbulent resuspension.

When space is a constraint, use of a special device such as a swirl-concentrator could be considered for solid separation. Warrer-Hansen (1982) reports that experiments with a swirl-concentrator were successful. Figure 5.8 illustrates the design of the device, which is a circular tank with a tangentially directed water flow, which introduces secondary motion flows that draw the solids to the bottom of the tank. An overflow weir at the surface in the centre ensures a clarified outflow. It is then possible to concentrate the solids and discharge them to a sewer using a flow of only 1–5% of the total. Though reported to be successful in initial tests, only very few appear to have been used under farm conditions, and then in conjunction with settling ponds. The space-saving advantage is, therefore, lost. Leffertstra and Kryvi (1988) reported that it is only moderately efficient. They found its requirement of a gentle flow of water, susceptibility to blocking and occasional problems in maintaining the correct vortex for maximum waste removal, important disadvantages.

A simple method of removing particulate matter from effluents that do not have high solid levels is by filtration through a single stationary mesh. When the solid content is high, the mesh may easily become clogged, requiring frequent cleaning. A self-cleaning or rotating filter may be an answer to this problem, but the rotating action of the screen may result in higher concentrations of dissolved nutrients in the discharged effluents.

A commercially available 'Triangelfilter' (trade name) is reported to filter the solids relatively quickly and thus reduce the leaching of nutrients into the remaining effluent. Mäkinen *et al.* (1988) have recommended it as a useful method to treat effluents from 'high-tech' aquaculture systems such as recirculation and heated water facilities. According to its manufacturer, Hydrotech of Sweden, it has a treating capacity of $15–35\,\mathrm{l\,s^{-1}}$. It is a surface pouring sieve which uses an inclined filter plate consisting of a plastic filter cloth sieve and a grid under it. The sieving cloth is mounted on an inclined plastic grid which creates barriers for the water pouring on the filter plate, increasing its filtering capacity. The effluent falling on the filter plate goes partly through it and partly on it, to reach the lower part of the plate, because the fine particles of the sediment gradually clog the meshwork in the sieving material. When the filter plate is fully clogged, it is flushed upwards automatically (with the activation of an electronic contact switch) by high-pressure water from moving nozzles. The sludge on the filter plate is transported to a sludge channel and piped into a settling tank.

Based on experimental work in a rainbow trout farm in Finland, Mäkinen *et al.* (1988) estimated the effect of the filter on phosphorus reduction to be somewhere between 80 and 40% in comparison with the swirl concentrator which usually reduces phosphorus loading from 60 to 10%. A settling tank is, however, necessary in conjunction with all such filter processes, as the collected sludge has to be held and removed.

Wastes from floating cage farms in the open seas or other water bodies generally settle below the cages and/or are flushed out by the currents. Several methods have been tried to reduce the accumulation of solid wastes under cages. Enell *et al.* (1984) found that most of the wastes could be collected by suspending a PVC funnel beneath the cages in a freshwater farm. However, it is not easy to maintain such devices beneath cages permanently without restricting the flow of water through the cages and stressing the fish stock. Their use in cages in exposed marine sites is likely to be much more difficult.

There are reports from Poland, Czechoslovakia and Sweden of experimental attempts to remove solid wastes from cages by suspending funnel-like devices. The wastes collected are pumped out at intervals. In Poland (Tucholski *et al.*, 1980a and 1980b), about 45% of the solid loadings could be removed, but only 15–20% of the total nitrogen and total phosphorus. In Czechoslovakian attempts (Hartman *et al.*, 1982), up to 40% of the solid material from a trout cage could be removed. The most successful experiment reported is from Sweden (Enell *et al.*, 1984), where 82% of the waste phosphorus could be removed, with higher efficiency obtained by removing sediments soon after feeding.

Submersible mixers have been used successfully to disperse wastes from sea-based cage farms (Braaten *et al.*, 1983) and to prevent further accumulation. By using the mixers, it was possible to reduce a 40-cm layer of accumulated waste to 10–15 cm within three months, and to prevent later accumulations. By suspending it above the seabed, the mixer could be used beneath the cages without having to move the fish from the site. However, it is necessary to take special care to prevent water that is low in oxygen being drawn into the cages, and to ensure that resuspension of sedimented organic waste does not increase BOD in water around the farm. Braaten *et al.* (1983) have also suggested the use of submersible pumps to collect and transfer sedimented waste to barges for further treatment.

Wray (1988) and Baklien (1989) describe prototypes of a floating raceway system that can use pumped water from different depths at sea and collect solid wastes in sumps. The wastes can be removed to barges using sludge pumps. Though waste dispersion into the sea can be minimal and the quality of the salmon grown in such a raceway is claimed to be better, the cost involved is likely to restrict its wider use.

From experience, it would appear that the only practical way of minimizing eutrophication in cage and pen farming is by siting the farms in locations that are well flushed so that solid wastes will be dispersed more easily. Where this is not feasible, it has been suggested that periodic redeployment of cages may reduce the effect of organic wastes on the benthic ecosystem. Use of single point moorings to obtain wider dispersal of wastes, or allowing the whole or sections of the farm to be fallow, as in agricultural practices, could help in reducing pollutive effects. The practice of rotating sites to allow disintegration and dispersal of wastes would also be applicable in the case of off-bottom culture of bivalve molluscs. As described later in Section 13.3, constructed wetlands growing macrophytes are being tried as a means of disposal of fish-farm wastes.

5.5.3 *Control of water-borne diseases*

Reference has been made in Chapter 3 to public-health concerns relating to certain types of aquaculture. This would appear to apply more to tropical aquaculture and especially to land-based stagnant pond farms as opposed to running-water ponds. Such stagnant water pond farms are generally built on low-lying swampy areas that are favoured places for mosquito breeding.

Human malaria infection, brought about by four different parasite species of the genus *Plasmodium*, is transmitted through about 60 species of *Anopheles* mosquitoes. As all carrier species of mosquitoes are aquatic breeders, the design and operation of pond farms should take into account the need for preventing mosquito breeding. Properly constructed farms, with adequate drainage facilities can minimize this risk. Shallow, weed-infested ponds without proper management and stagnant puddles become as inviting to mosquito breeding as the swamps on which the ponds were built.

A depth of at least 60 cm and preferably 1 m, and steep inside dike slopes to avoid shallow margins, are recommended to reduce suitable sites for mosquito breeding. The ponds have to be kept free from all weeds, particularly floating weeds, and all emergent vegetation cut back. Any seepage from the ponds has to be drained off through seepage channels. Cattle should be prevented from grazing on pond embankments as hoof prints are well-known breeding grounds for mosquitoes in the humid tropics.

The problem of mosquito breeding is much less prevalent in coastal ponds regularly flushed with tidal water, even though they may be comparatively shallow. However, it often happens that large mats of benthic algal complex, which generally grow in such ponds, become dislodged from the bottom and accumulate on pond margins or corners serving as substrates for mosquito breeding. The disintegrating algae can produce unpleasant odours and become a public nuisance, which can be avoided by removing the floating algal complex regularly and composting it for use as fertilizer.

Adults of some species of fish, and larvae of most species, feed on mosquito larvae, and many on the algal vegetation that shelters them. Stocking ponds with larvicidal fish, such as *Haplocheilus panchax* and *Gambusia affinis*, has also been found to be helpful.

The other important water-borne disease associated with aquaculture in land-based farms is bilharzia or schistosomiasis. Human schistosomiasis is caused by at least three species of bloodflukes (trematodes), namely *Schistosoma haematobium* (genito-urinary bilharziasis), *S. mansoni* and *S. japonicum* (intestinal bilharziasis). The adult forms mature in the blood of human or other hosts, and the eggs laid in the plexus around the colon escape through the bladder or intestinal wall into the urine or faeces. On contact with water, the eggs hatch and produce free-swimming larval forms called miracidia, which must find suitable snail hosts within a day or perish. On finding the intermediate host, a miracidium penetrates its skin and establishes itself and multiplies to form further free-living larval forms called cercaria. These emerge from the snail and swim about until they find a human host. On contact, they bore through the unbroken skin of the host (or through the buccal mucosa if the water is swallowed), make their way to the liver, mature and mate, lay eggs and continue the cycle.

It is evident that fish ponds in endemic areas can create conditions suitable for the transmission of the disease if they are contaminated by infected people and have the necessary environmental conditions for growth of the snail populations that serve as intermediate hosts.

A two-pronged approach to prevent fish farms becoming sources of infection are (1) by avoiding as far as possible contaminated water as a source of water supply, and enforcing adherence to personal hygiene by farmhands and the neighbouring community, and (2) by the control of snail infestation.

There are several genera of water snails that are known to be vectors. Clearance of emergent and submerged vegetation can greatly help in controlling snail infestations. Monoculture of snail-feeding fish, or inclusion of such fish in polyculture combinations, would help in the control of snail populations. There are also smaller snail-feeding fish, such as *Astatereochromis alluaudi* and *Haplochromus* spp., which could be introduced into ponds for the sole purpose of controlling snails. Integrated fish and duck farming, practised in many areas, serves as an effective means of control, as ducks feed on snails as well as on algal vegetation in ponds.

Periodic draining and drying of ponds, followed by liming of the pond bottom, is considered to be the most reliable means of avoiding snail infestation. Parasitized snails are generally unable to survive prolonged desiccation, and periodic drying of the pond is an effective means of eliminating snails carrying *Schistosoma*.

In addition to the hazard of spreading water-borne diseases, which can be minimized by improved design and operation of farms, there are also other public health problems that can be created by some aquaculture practices. These include the use of large quantities of animal manures, the use of waste water including domestic sewage, and the growing of filter-feeding molluscs in contaminated waters. These will be discussed in succeeding chapters. However, one practice relevant to farm design is integrated farming of fish with ducks and pigs, where special design features may be required to prevent unsanitary conditions developing in the farms. These may include special fermentation tanks for preliminary treatment of wastes from pig sties, or duck houses over ponds to allow direct manuring in small doses.

Chapter 6
Use of Natural Resources

6.1 Sources and Utilization of Land Water Resources

It is estimated that in semi-intensive aquaculture 11 000–215 000 m³ of water is required for the production of one tonne of fish, and in intensive farming the requirement may be about 29 000–43 000 m³. Intensive forms of flow-through aquaculture systems use comparatively large quantities of water. In well-managed ponds, the mean water supply can range between 8 and 25 l s⁻¹ per ton of annual production. Production farms generally depend on surface and spring water, most of which flows back to the waterways as residual waste water. In stagnant-water pond farms, water use is generally limited to the initial filling to the required depth, and subsequent topping up to make up for loss due to evaporation and seepage. For a fish pond with an average depth of 1.5 m the amount of water required to fill it initially is about 15 000 m³ ha⁻¹ (Pillay, 1990). Loss of water through evaporation and seepage varies depending on the climate and soil conditions. In Europe, the loss is reported to be about 0.4–0.8 cm d⁻¹, whereas in arid tropical regions it could amount to 1–20 cm d⁻¹ or more.

In intensive culture daily water exchange may be 500%, in addition to vigorous flushing of sludge, in order to bring water quality almost to that of input water. As waste water production has an important role in environmental sustainability, water exchange rate is minimized in modern aquaculture practices (Avnilmech, 1998). Nevertheless, user conflicts can arise when sources of water supply are irrigation systems, and aquaculture is given less importance than agriculture. Conflicts can also arise with downstream uses, when streams or rivers are contaminated by effluent discharges from farms, especially when the farms are infected with pathogenic organisms and the effluents are not treated adequately before discharge.

Comparatively less water is used in a hatchery, and the preference is generally for spring or borehole water to reduce the chances of contamination. When the same sources are used for other purposes, it is necessary to determine the long-term sustainability of multiple uses based on quantitative estimates of the water resources, including their renewal, utilization and recycling.

Coastal land-based farms often depend on diurnal tides for filling as well as draining. When the tidal range is not suitable for the purpose, pumping of marine or brackish water may be resorted to. This seldom creates any conflicts with other users, except in certain areas where the saline water from the ponds has been found to penetrate the soil of agricultural land.

However, conflicts have arisen in recent years when large quantities of fresh water are pumped by shrimp farmers to reduce salinity in coastal ponds. It is reported that the settling of the overlying land as a result of such intensive pumping has resulted in increased hazard of seasonal flooding. Abstraction of large quantities of underground fresh water also results in the infiltration of saline water into aquifers, making the water unsuitable for agriculture or human consumption.

6.2   Use of Animal Wastes

Animal wastes are used to fertilize pond farms in many countries of the world and are considered superior to inorganic fertilizers in producing and maintaining desirable species of planktonic and benthic food organisms in fresh and brackish water ponds.

In integrated livestock-cum-fish farming, the animals are raised near or on fish ponds so that the manure and other waste materials can be discharged directly into the ponds. The number of animals is regulated to ensure that the wastes discharged into the ponds do not exceed the quantities that the biological processes can handle without creating adverse environmental effects within the ponds.

Total solid contents of animal manures are high; for example, for pigs and poultry it is reported to be 268–509 g (dry weight) d^{-1} per animal (Whetstone *et al.*, 1974, quoted by Nash & Brown, 1980). According to the data provided by Woynarovich (1980), the solid contents of duck manure works out to 50.5 g (wet weight) d^{-1} per duck.

The number of animals raised in association with fish farms is generally based on the extent to which the pond waters decompose and mineralize these solids and use the nutrients for plankton or benthic production. The solids settle at the pond bottom and the easily decomposible constituents undergo anaerobic decomposition, producing carbon dioxide, ammonia and hydrogen sulphide. Then direct aerobic decomposition takes place using oxygen in the water and the atmosphere, producing carbon dioxide, nitrite nitrogen and sulphur. These are utilized by algae to grow and reproduce, and produce oxygen by photosynthesis. This oxygen, along with that dissolved in the water from the atmosphere, enhances the aerobic decomposition of the original wastes, as well as the new wastes received in the system.

Schroeder and Hepher (1979) confirmed experimentally that the non-mineralized fraction of the manure is used as a food base by bacteria and protozoa. Zooplankton seem to ingest small manure particles and digest the bacteria that coat the particles. The residual particles are then expelled.

After the application of manures or chemical fertilizers, the pond water is allowed to stand for an appropriate period to ensure that they are mineralized and dissolved, or at least the solid matter is settled on the pond bed. The organic load of the effluents discharged from the ponds is thus generally minimal. From experimental work in Hawaii, Lee *et al.* (1986) found that nutrient levels of discharges from shrimp ponds manured at weekly intervals (with no supplemental feeding) were less than or equal to nutrients in the in-coming water. Based on this, they concluded that 'a marine shrimp pond can be considered a dissolved nutrient marine treatment plant converting unwanted cattle manure . . . into a valuable commodity—shrimp'. Even in high-density culture of the catfish *Clarias* sp. in Thailand, where the fish are fed with large quantities of trash fish, the percentage of undecomposed organic matter in the effluents discharged into the environment is not significantly high. After harvesting, the farmer removes excess undecomposed organic matter which has settled on the pond bottom and uses it as fertilizer for his crops.

In integrated fish–animal production, many farmers now do not apply raw animal wastes in the ponds, but allow the solids to settle and ferment in specially constructed fermentation tanks, and allow only the supernatant clear liquid wastes into the ponds. This reduces the load of solids in the ponds as well as bacterial populations, which may include pathogenic organisms. The sludge from these tanks is used for fertilizing terrestrial crops.

It has been speculated that livestock–fish systems could cause the development of influenza pandemics (Schotissek & Naylor, 1988) but there appears to be no supporting evidence for this hypothesis.

6.3   Use of Sewage

The use of domestic waste water and human sewage for fish farming is an age-old practice in some Asian countries and has been adopted in some others, though not so widely. Use of human sewage is not basically different from the application of animal manure and wastes, but there are serious constraints to its use, largely for aesthetic reasons and the possibility of human disease transmission. Nevertheless, considerable experience has accumulated to show that fish ponds can have a role in sewage treatment.

Efficiency of oxidation ponds used for aerobic decomposition of human wastes has been found to be enhanced through combining fish culture with waste treatment. The biological processes involved are similar to those already described for animal wastes. One of the major environmental problems in waste treatment in oxidation ponds is the production of algal blooms which increase BOD values in the effluents. Further, the nutrients concentrated by the algae have to be removed from the effluent to prevent eutrophication of the receiving water and to meet the discharge standards set up by regulatory agencies. Culture of phytophagous fish is an efficient way of reducing algal cells from oxidation pond water.

By proper management, the risks of the ponds turning septic and the treatment becoming ineffective can be avoided. High concentrations of calcium, high alkalinity and pH values in fish ponds are favourable conditions for the removal of nutrients that cause eutrophication of receiving waters (Hepher, 1958; Rohlich, 1964). Large amounts of phosphorus are bound up in pond muds, plants and plant-feeding organisms. Most of the nitrogen under high pH conditions is transformed into gaseous NH_3 and escapes to the atmosphere. Nitrates are taken up by plants, and their removal by growing phytophagous fish is an effective means of improving effluent quality and, thereby, reducing eutrophication in receiving waters.

The quality of effluents from a rearing facility depends on the water retention time. In facilities fertilized by animal wastes or human sewage, outflow of water is very limited. Except in emergencies, effluents are discharged only at the end of the crop season. Even when some outflow is allowed, the retention time is not likely to be less than a month, which gives adequate time for biological decomposition and utilization of wastes. It should, therefore, be easier to meet discharge standards and prevent adverse environmental impacts.

Schroeder (1975) observes that fish in waste-water ponds have the effect of improving oxygen regime and enhancing pH as a result of high photosynthetic rates during the day, though the values may go down at night. Increased dissolved oxygen and pH increase the rate of disinfection of coliforms (Oswald, 1973).

None of the available literature has indicated any proven evidence of human bacterial diseases being transmitted through fish culture using animal wastes or sewage. Studies such as that of Carpenter *et al.* (1974) showed virtually no human pathogens in cultured fish, and a drastic reduction in coliforms. In experimental studies, Allen *et al.* (1979) found faecal streptococci and other organisms associated with human diseases (*Aeromonas*, *Pseudomonas*, *Klebsiellae*, *Salmonellae*, *Escherichiae*) only in gut contents of small Pacific salmon and rainbow trout grown in domestic waste-water-fertilized ponds, and none in other visceral organs or musculature.

Wild fish in infected or endemic areas have been known to harbour enterobacteriaceae that cause disease in humans or other warm-blooded animals. Vibrios causing epidemics such as cholera have been isolated for varying periods in experimentally infected fish, after which they lose their virulence (Pillay *et al.*, 1954). But there is no evidence that any of the pathogens have become systemic under fish culture conditions. According to Allen and Hepher (1969), most of the epidemics attributed to waste water sources are from raw sewage gaining access to food eaten directly by man, or from contamination of water supply systems by untreated sewage. Recent studies, in which treated sewage was used to grow fish biomass for predatory fish (Edwards *et al.*, 1987), showed that both the predatory fish and the forage fish were safe for human consumption.

Depuration of fish reared in waste waters, by keeping them in clean water for a number of days before marketing (as is now done in many countries), would be a primary precautionary measure against transmission of human diseases from waste-water-reared fish. An additional precaution is to avoid eating raw or improperly processed fish or shellfish grown in waste-fed ponds.

6.4   Use of Heated-water Effluents

The discharge of large quantities of heated-water effluents, containing circulated condenser-cooling water, from power plants into natural water bodies is often detrimental to the environment. Treatment to make them harmless, as in the case of all other waste treatment, is expensive and may require extensive facilities. However, in temperate climates the heated effluents can be used beneficially in aquaculture to promote the growth of fish and shellfish all through the year, and enable the culture of organisms that require higher temperatures than are normally available in such climates.

In practice, there have been major problems in developing economically viable aquaculture enterprises based on the use of heated effluents from power stations, even though in theory it should be an ideal way of combining waste disposal with food production. Experiments are underway to develop economical re-use systems (Lasordo *et al.*, 2001; FFI, 2002; Timmans & Lasordo, 1994) either to reclaim waste water or to produce heating water for fish farming. The quantities of warm water discharged from the condensers of a modern electricity station adopting direct cooling are much more than can be utilized by even a very large farm. Aston (1981) calculates that the 'reliably available' water (allowing for plant breakdown and repair) of a modern direct-cooled 2000 MW station ($15.8 \text{ m}^3 \text{ s}^{-1}$) would be 20 times the requirement of a modern 200 tonne eel farm. So a fish farm alone may not always be able to solve the problem of disposal of cooling water from power stations. Also, the continuity of heated water flow cannot be relied upon as it is dependent on the continuity of electricity generation.

Other risks related to the use of heated-water effluents are the water quality changes that occur as the condenser cooling-water circulates through the power plant system. In normal power-plant maintenance, biocides such as chlorine are used to inhibit biofouling in the condenser tubes and in the water treatment system. Chlorine may be injected intermittently or on a continuous basis to give a free residual concentration upon discharge of $0.2–1.0 \text{ mg l}^{-1}$. The Environmental Protection Agency of the USA limits chlorine concentrations in plant discharges to a maximum of 0.5 mg l^{-1} and an average of 0.2 mg l^{-1} free available chlorine for up to two hours per day (Nash & Paulsen, 1981).

Besides a number of other biocides, chemicals are also used for corrosion inhibition and for pH control, or as dispersing agents in cooling towers. There is always the likelihood of residuals of these chemicals in the effluents. In addition, cleaning agents and cleaning products of boiler blow-down and maintenance are discharged into the cooling water system. Cooling water blow-down may be intermittent or continuous, but constitutes the largest effluent component in most plants. A number of other miscellaneous wastes also normally find their way into the system.

All the consequent water quality changes, though potentially harmful, have not affected the use of the effluents in aquaculture, probably because of the treatments that many plants undertake to neutralize the waste and the substantial dilution it undergoes in receiving waters. The major constraint appears to be the problem of synchronizing the volume and temperature of effluents with the requirements of aquaculture under specific circumstances.

Despite the limitations indicated, heated water effluents are used for small-scale aquaculture in certain areas. They are also used for raising the water temperature in small lakes to enhance fish production. Retention of the effluents in farms, which results in a reduction in temperature and removal of toxic substances, would help to eliminate the harmful effects of direct discharge into water bodies.

6.5   Recycling of Water

From the point of view of minimum environmental impact, recycling or recirculation systems of aquaculture would be ideal, as there would be only reduced discharges into open water bodies (Lasordo *et al.*, 2001). Water leaving the rearing facilities is returned to the system completely or partially by pumping. Obviously, the use of water from natural sources will also be reduced significantly. In simple systems of water recycling, the main treatment may be aeration or oxygen injection, or mechanical filtration for removal of solids. When the species being cultivated can tolerate relatively high concentrations of metabolic by-products, simple systems of recycling can be adopted. However, when the species is sensitive to such wastes and the quantity of make-up water available is limited, additional water treatment facilities would also be required.

In completely closed systems, water is supplied only to fill the system and replace evaporation. There is very little discharge of waste water, as the same water is recycled (in theory indefinitely) after appropriate treatment, which may include among others, sedimentation, mechanical or biological filtration, sterilization, oxygenation, aeration, degassing, cooling or heating and pH control.

Despite several design and operational problems, a number of experimental and commercial projects using recycled water have been undertaken in industrialized countries. Experience so far seems to indicate that much more research and development work is required to make the system economically viable and to achieve the objective of cutting down water usage to the extent originally conceived. However, it is obvious that when the technology is adequately advanced and applied, it could contribute substantially to reducing adverse impacts of intensive aquaculture. Even now, the present technology can be employed successfully in hatcheries for production of fingerlings, with appropriate management.

6.6   Use of Trophic Levels in Aquaculture

Polyculture of aquatic species is considered to be an environmentally benign system when different trophic levels in a waterbody can be utilized for rearing ecologically compatible species without competing for living space and food resources available. In traditional fish farming, generally herbivorous and omnivorous species occupying different water levels are used. However, it has been noticed that some of the low trophic level feeders can also be highly selective in their feeding habits, as in the case of filter-feeders that require plankton of a particular size or shape. As will be indicated below, compatibility may be limited to fry or fingerling stages only. Similarly, species that are hardy and tolerant of unfavorable conditions will have the advantage of better survival in relatively poor environmental conditions.

Among the species used in polyculture of trophic level feeders are a group of five Chinese carp, the grass carp (*Ctenopharyngodon idella*), the silver carp (*Hypophthalmichthys molitrix*), the bighead (*Aristichthys nobilis*), the black carp (*Mylopharyngodon piceus*), and the mud carp (*Cirrihina molitorella*). Experience showed the compatibility of the adults of these species in ponds and the higher production the farmers could get by combined culture. The grass carp is herbivorous and feeds on macrovegetation, including grass and aquatic plants; the silver carp feeds on plankton, mainly phytoplankton; the bighead consumes macroplankton; the black carp feeds on snails and other molluscs at the bottom of ponds. The mud carp feeds primarily on detritus on the pond bed. To this combination the omnivorous common carp (*Cyprinus carpio*) is added to obtain higher yields.

Indian major carp (*Catla catla*) is a surface and column feeder. Young fry feed on planktonic unicellular algae. Fingerlings start feeding on zooplanktonic organisms. Adults feed on different types of algae, planktonic protozoa, rotifers, crustaceans, molluscs and decayed macrovegetation. Rohu (*Labeo rohita*), another Indian major carp, is a column feeder. Larvae and young fry feed on unicellular algae and zooplanktonic organisms. Adults feed on various types of vegetable matter, including decaying aquatic plants, and algae. The adult mrigal, a major Indian carp, is a bottom feeder. Calbasu, another major Indian carp (*Labeo calbasu*), is also a bottom feeder, consuming benthic and epiphytic organisms and organic debris. Indian farmers try to combine surface-, column- and bottom-feeding fish or their fingerlings with the Chinese silver carp and grass carp, together with the common carp, to fully utilize the ecological niche according to local conditions.

Chapter 7
Waste Production in Aquaculture

To assess the environmental impact of aquaculture effluents, it is necessary to know the background concentration of nutrients in the receiving waters, as well as the emission of nutrients from the farm per unit of time, and the retention time of water in a given area. Though the types of wastes produced in aquaculture farms are basically similar, there are differences in the quality and quantity of the components depending on the species cultured and the culture practices adopted. Much of the available information relates to intensive systems of salmonid culture and, to a large extent, to pens and cages in temperate climates. Even though the processes involved may not differ significantly, the applicability of such information to other forms of aquaculture in different climatic areas cannot be taken for granted. However, it could provide the background to future investigations and the basis for any *ad hoc* assessments.

7.1 Feed-derived and Metabolic Waste Products

The main types of waste in hatcheries or production farms can be described as: (1) residual food and faecal matter, (2) metabolic by-products and (3) residues of biocides and biostats. In land-based farms using fertilizers for the production of food organisms, wastes can be built up as a result of their sedimentation, disintegration and under-utilization. In tidal water fish ponds, the inflows may bring in additional organic matter in the form of suspended solids and dissolved nutrients. In intensively fed farms, the composition and physical nature of the wastes reflect the composition of the diet and the digestibility of its components.

Feed is generally provided in dry, moist or wet form. Dry and moist pellets are more commonly used in salmonid culture. There are significant differences in the composition of commercially available feeds, but all contain protein, carbohydrate and fat, besides some additives such as vitamins, therapeutants and pigments. The percentage of the components as well as the energy contents are generally declared on the labels. Håkanson *et al.* (1988) give typical values of feed composition in

Table 7.1 Composition of three types of typical salmon feeds used in Nordic countries.

	Moist pellets	Dry feed (low energy)	Dry feed (high energy)
Dry matter (g kg^{-1})	325	900	900
Protein (g kg^{-1})	170	500	450
Fat (g kg^{-1})	60	120	240
Carbohydrate (g kg^{-1})	50	150	100
Nitrogen (g kg^{-1})	27	80	72
Phosphorus (g kg^{-1})	4	15	10
Gross energy (Mcal kg^{-1})	1.3	4.6	5.2

(From: Håkanson, 1988)

salmon feeds used in Nordic countries (see Table 7.1). All the feed components, and also the by-products of metabolism, could become waste products. These products would contain organic carbon and organic nitrogen (carbohydrate, lipid and protein), ammonium, urea, bicarbonate, phosphate, vitamins, therapeutants and pigments. The more important components of waste food and faeces are organic carbon and nitrogen compounds, which sink to the sediment.

Food and feeding are probably the most important factors involved in the management of aquaculture from the environmental and economic points of view. Modern aquaculture is based mostly on feeding of manufactured feeds, except the culture of bivalves and seaweeds, as described in Chapter 8. The content of phosphorus and nitrogen in the feed and the feed conversion rates are of importance in assessing environmental impacts of aquaculture in any related project (Ackefors, 1999). The feed coefficient in many North European countries has decreased from 2.3 to less than 1.3 as a result of improved knowledge. The nitrogen content of commercial feeds has been decreased from 7.8%, and the phosphorus content from 1.7%, to <1.0%. As a result, for every tonne of fish produced, the discharge of phosphorus in now <10 kg, and of nitrogen <53 kg (Ackefors & Enell, 1994).

Processing methods adopted in commercial feeds are of importance in reducing pollutive effects of feed-derived wastes. Extruded pellets have a slow sinking rate and higher water stability and, therefore, become available to fish more effectively than ordinary compressed pellets. The ingredients that the feeds are composed of are also important from the point of view of waste production. Commercial salmon feeds now have a composition of about 30% fat, 40% protein and 13% carbohydrate, with an energy content of 19.2 MJ kg^{-1}. The nitrogen content is now about 7%, and the fish utilizes fat instead of protein for energy and smaller volumes of nitrogenous compounds such as ammonia are excreted. There is less excretion of phosphorus since it has been reduced to about 1% in the diet.

Feed management includes the size of feed and the regulation of size of biomass and age composition, as well as intervals of feeding according to environmental conditions. Many advanced farms, whether land-based or offshore, hatcheries or rearing facilities, use computer programmes to regulate feeding according to daily variations in the parameters to avoid waste and feed spillage. By the use of such measures and adjusted feeding procedures, feed conversion ratios have been increased, and the quality of farm effluents discharged into the waterways enhanced (Pillay, 1999).

Commercial fish feeds generally contain fish meal as a major component, even in improved formulations (Enell, 1995), in order to reduce waste discharge. In the light of the controversial and predicted shortage of fish-meal and alleged over formulation (de Silva, 1999), the search for suitable substitutes has to be continued. Manufacturers, besides reducing fish-meal as the source of protein, use meat-meal, bone-meal, blood-meal, poultry-by-product-meal and dried brewer's yeast as substitutes for fish-meal. Kaushik *et al.* (1995) reported on the effect of partial or total replacement by soybean protein on growth, protein utilization, potential oestrogenic or antigenic effects, chlorostromia and fish quality. Donaldson (1997) states that if aquaculture should continue to grow, alternate protein sources other than fish-meal should be studied. Plant proteins offer an alternate source that has been tried in the culture of several species of fish. Processed soya-bean-meal in catfish, alfa-leaf protein and canola-meal in tilapia, and processed canola protein in rainbow trout have been tried. Irrespective of whether alternate sources have proved suitable for partial or complete replacement of fish meal, it is certainly possible to reduce the percentage composition of fish-meal in fish and other animal feeds.

Wilson (1994) has reported that there is a general consensus that fish have relatively poor ability to utilize carbohydrates as an energy source, but this view is said to be changing. Many fish-feed technologists are optimistic about the commercialized use of other sources of protein, like single-cell proteins and leaf concentrates. It is expected that problems of heat-stable alkaloids, cystine, anti-nutritional elements, and deficiency of methionine, etc., can be solved by suitable research. It is well recognized that in many of the cultured species, a lower protein diet could be utilized without compromising growth and production. It is expected that the aquaculture industry is likely to develop further in the tropics, where there are may instances of effective use of lower protein diets (Tacon, 1996; Tacon & de Silva, 1994; de Silva, 1999).

The production of faecal and excretory wastes obviously depends on the stock density in farms. Experimental studies with rainbow trout have shown that the quantity of solid and dissolved organic metabolic products is linearly related to the quantity of food absorbed (Butz & Vens-Cappell, 1982). The faecal production was calculated to be around 260 g dry weight per kg of food consumed, which works out to 26% of the food consumed. Rychly & Spannhof (1979) found high protein digestibility in rainbow trout for all experimental diets, but the digestibility of diets containing less protein and a higher percentage of carbohydrates was comparatively less. (The digestion coefficient of the whole diet appears to be in general agreement with the findings of Butz & Vens-Cappell (1982).) The digestibility of the components governs the composition of the faeces and urine. Rychly (1980) found that the composition of the urine is influenced by the protein content of the food consumed by the fish. Data presented by Penczak *et al.* (1982) suggest that rainbow trout faeces contain 30% carbon, 4% nitrogen and 2% phosphorus.

Data on the protein content of the food consumed and of protein digestibility provide an estimate of the excretory nitrogen produced in a farm. Based on the nitrogen content of salmon (2.72%) and of the feed normally used for rearing the species (7.68%), Gowen & Bradbury (1987) calculated that of 122.9 kg of nitrogen consumed, only 27.2 kg are retained, assuming a food conversion ratio of 2:1. In other words, 22% of the consumed nitrogen is retained and 78% lost as faecal and excretory nitrogen. Penczak *et al.* (1982) found a nitrogen retention of 30% in rainbow trout. If the nitrogen content of the faeces is 4% (see Penczak *et al.*, 1982), 68% to 86% of the consumed nitrogen is voided from the fish as soluble ammonium and urea, which approximates to a production of 32 kg of ammonium per tonne of food fed (Gowen & Bradbury, 1987). Bergheim *et al.* (1984) found that ammonium in effluents exceeded 100 times the quantity supplied in the feed, which supports the view that ammonium is mainly an excretion product. Ammonium production in salmonid hatcheries has been reported to be 28.9 kg (Liao & Mayo, 1974) and 32 kg (Willoughby *et al.*, 1972) per tonne of food consumed. In land-based trout farms, ammonium excreted per tonne of fish is estimated to be 45 kg in Denmark (Warrer-Hansen, 1982), and 55.5 kg in the UK (Solbe, 1982). As most of the excretory nitrogen consists of ammonium, and the utilization of ammonium and urea within the farm and in the receiving waters is for phytoplankton growth, it has been suggested that these need not be considered separately.

Bicarbonate is a by-product of respiration and is excreted via the gills. Even though it is alkaline, it is unlikely to affect the pH of seawater in any significant way, because of the buffering action of seawater.

Phosphates produced in the farm are in particulate form, as against nitrogen which is mostly in dissolved form, and are reported to have minimal environmental effect in marine waters. Even though there are differences of opinion on this, many workers are of the view that except in some low-salinity areas, phosphate is not important in controlling algal growth in coastal waters and should not therefore be considered as an important waste product. The importance of both nitrogen and phosphorus as waste products is emphasized by others (see Håkanson *et al.*, 1988). In freshwaters, enhanced levels of dissolved inorganic phosphate have been found to cause eutrophic conditions. Bergheim *et al.* (1984) found increased phosphorus concentrations in a freshwater salmonid farm with about 60% soluble PO_4–P, 28% soluble polymerized and organic phosphorus, and 12% particulate phosphorus. About one-half of the total phosphorus supplied in the feed was lost through the outlet of the farm.

Waste production per unit of fish biomass was found to be related to fish size. The build-up of wastes derived from feed and metabolism is subject to seasonal variations, depending on the climatic conditions and especially water temperature. Under favourable climatic conditions feeding is possible throughout the year. In areas with severe winter conditions, feeding rates are drastically reduced to survival level. This has to be taken into account when estimating the potential waste production in a farm.

7.2   Wastes from Food and Feedstuffs

The factors involved in the production of wastes by the stocks feeding on natural food produced in the farm, or unprocessed feedstuffs fed to them, are very similar to those already described in Section 7.1 for processed feeds. For example, the accumulation of faecal matter and detritus under a raft in a mussel farm can be very substantial. According to Figueras (1989), an individual mussel can filter between 2 and 5 litres of water in an hour, a rope of mussels can filter more than 90 000 litres of water in a day, and a raft of mussels can filter 70 million litres of water a day. The retention percentages of plankton and detritus as a result of filter-feeding are estimated to be 35–40%. Approximately 180 tonnes of organic matter is, therefore, ingested by a raft of mussels, of which 100 tonnes are returned to the sea. This adds to the sediments to be decomposed at the sea-bottom in the mussel beds.

Considering the density of populations in intensive culture, the accumulation of faeces and pseudofaeces can be extremely high in all types of bivalve culture. Arakawa *et al.* (1971) estimated the output of faeces and pseudofaeces from a typical oyster raft in Hiroshima Bay, Japan, holding 420 000 oysters over a nine-month period, as 16 tons. With about 1000 rafts under operation in the bay (Arakawa, 1973), the sediment deposition can be very significant. Dahlbäck and Gunnarsson (1981) studied the sedimentation and dissimilatory sulphate reduction under mussels reared on long lines in Sweden. They found deposition of over 1000 g organic carbon m^{-2} per year, reflecting the rate of production of faecal matter. This was three times more sedimentation than in areas further away from the mussel farm.

Phaeopigments produced by the conversion of chlorophyll *a* as a result of the mussels grazing on phytoplankton were found to deposit at a rate of 10 times more under the mussels than in areas far from the farm. Ottman and Sornin (1985) report that the organic carbon production by oysters grown on racks in France varies seasonally from 7.6 to 99 g m^{-2} d^{-1}, correlated with seasonal variations in phytoplankton. The bacterial decomposition of the organic matter results in high rates of consumption of O_2, often creating a reducing environment and the production of H_2S, which is toxic to oysters (Ito & Imai, 1955; Caldwall, 1975).

Solberg & Bregnballe (1982) studied the effects of wastes from farmed trout fed with minced trash fish in Denmark and found significant increases in BOD, total phosphorus as phosphate and orthophosphate, ammonia and organic nitrogen, four to six hours after feeding. These are attributed to waste food particles and oil droplets. After this period, the waste output settled at a 'basal' level. Contrary to expectations, they found substantial discharges from farms even when there was no feeding, and these may be due to stirring up of sedimented faeces and small feed particles by movements of the fish.

7.3   Feed Loss

Feed-derived wastes in a farm include not only faecal and other excretory products, but also unconsumed feed. Feed loss depends on a number of factors, including the feeding behaviour of the stock, water stability of the feeds, the method of feed distribution and the time of feeding. Warrer-Hansen (1982) reported the percentage of unconsumed feed in Danish trout farms as: trash fish 10–30%, moist pellets 5–10% and dry feed 1–5%. These figures appear to be based on observations in pond and tank farms. Feed losses in marine or freshwater cage farms are likely to be higher. Braaten *et al.* (1983) estimate a loss of about 20% in marine cage farming of salmon. Feed loss in a land-based shrimp farm is reported to be over 10%.

In a comparative study of food conversion ratios of trout in freshwater ponds and cages, Beveridge (1984) found a higher ratio of approximately 20% in cage-reared fish, which was assumed to be due to higher rates of feed loss. Based on the data presented by Penczak *et al.* (1982), Gowen & Bradbury (1987) calculated a wastage of 27% and 31% for dry and moist diets, respectively, in freshwater. The unconsumed feed settles at the bottom of the farm. It is reported that in cage farms at sea, no substantial loss of carbon or nitrogen is likely by solution or microbial activity. However, in a land-based fish farm there is every likelihood of its disintegration as a result of the activity of the fish stock and aeration. For example, even though there is a loss of over 10% of the feed in a shrimp pond farm, under normal operating conditions no major increases are observed in the nutrient concentration in the effluents, except during final harvesting and pond cleaning.

7.4   Methods of Measuring Waste Production

Obviously, it is essential to have as accurate an estimate as possible of the quantities of polluting agents in wastes produced in a farm and

Table 7.2 Basic composition of feed and fish for drawing up a mass balance.

	Feed	Fish
Protein (g kg^{-1})	500	175
Fat (g kg^{-1})	200	180
Carbohydrates (g kg^{-1})	120	–
Nitrogen (g N kg^{-1})	80	28
Phosphorus (g P kg^{-1})	10	4.5
Gross energy (Mcal kg^{-1})	5.2	2.7
BOD (g O$_2$ kg^{-1})	1680	800

(From: Håkanson *et al.*, 1988)

discharged into the receiving waters, so as to reduce environmental impacts. However, there are major practical difficulties in conducting the necessary field measurements. Håkanson *et al.* (1988) have proposed a comparatively easy means of determination, based on the fact that production of the major polluting agents, namely, organic matter, nitrogen and phosphorus, is equal to the difference between what is added by the feed and what is utilized for fish production. This method should be applicable in intensive culture systems based on complete feeding, as in cages, raceways and tanks. The load of wastes from a farm or production unit can be calculated based on data on the composition of fish, feed, feed consumption and fish production or growth. The quantities of (a) organic matter, nitrogen and phosphorus lost through the faeces and (b) excretions of carbon dioxide and ammonia through the gills can be based on the metabolizability of the feed.

An example of the calculation given by Håkanson *et al.* (1988) is summarized below. The calculation is based on a commonly used feed in Danish sea farms containing 50% protein, 20% fat, 12% carbohydrate and 3.8 Mcal metabolizable energy. The fish reared is rainbow trout and the feed conversion ratio is 1.5 (dry feed); the digestibility of protein and energy is 85% and 81%, respectively. The compositions of the feed and the fish are given in Table 7.2, and form the basis for calculating a mass balance. Since the feed loss is included in the estimation of feed conversion ratio, it is not considered separately. Though it will not affect the total load of wastes, it will affect the distribution on the different parameters in Table 7.3, showing the mass balance.

Table 7.3 shows the calculation of waste load resulting from the production of 1 kg of fish. The quantity of organic matter is expressed as energy and BOD. It can be seen from the table that 35% of the energy is utilized for fish growth, 47% for respiration and the remaining amount goes as faeces (18%). The fish incorporates about 25% of the nitrogen, 15% is lost in faeces and 60% is excreted as ammonia from the gills.

Table 7.4 presents the mass balance based on a daily growth rate of 0.007 kg. It can be seen that 1 kg of trout growing at a rate of 0.007 kg d^{-1} fed on 55 kcal d^{-1} (i.e. about 1.1% dry feed d^{-1}) will consume

Table 7.3 Mass balance for energy, organic matter (BOD), nitrogen and phosphorus from the production of 1 kg of fish.

	Energy (kcal kg^{-1})	BOD (g kg^{-1})	Nitrogen (g N kg^{-1})	Phosphorus (g P kg^{-1})
Feed	7829	2416	120.0	15.0
Fish	2746	848	29.6	4.5
Faeces	1439	444	18.0	10.5
Excretion, ammonia	430	133	72.4	–
Respiration	3213	992	–	–
Waste load: faeces + excretion	1869	577	90.4	10.5

(From: Håkanson *et al.*, 1988)

Table 7.4 Mass balance presented as rates. Based on data from Table 7.3 and a daily growth rate of 0.007 kg.

	Energy (Kcal kg^{-1} d^{-1})	BOD (g kg^{-1} d^{-1})	Nitrogen (g N kg^{-1} d^{-1})	Phosphorus (g P kg^{-1} d^{-1})
Feed offered	54.8	16.9	0.84	0.11
Growth	19.2	5.9	0.21	0.03
Faeces	10.1	3.1	0.13	0.07
Excretion, ammonia	3.0	0.9	0.51	–
Respiration	22.5	6.9	0.51	–
Waste load: faeces + excretion	13.1	4.0	0.63	0.07

(From: Håkanson *et al.*, 1988)

6.9 g O_2 d^{-1}. The discharge will be 4.0 g BOD, 0.63 g N and 0.07 g P d^{-1}. The relation between respiration and BOD in the faeces is not precise, but it is not considered important as the BOD in the faeces is generally degraded quickly.

7.5 Fertilizer-derived Wastes

Most of the studies made on waste production and effluent qualities relate to intensive flow-through systems. The main focus of attention has, therefore, been on feeds and faeces. However, a substantive part of aquaculture in tropical and sub-tropical areas is in pond farms where the application of organic or inorganic fertilizers is either the sole or partial means of food production for the cultivated stock.

The use of animal wastes and human sewage has been described in Chapter 6, dealing with waste water use. The process of decomposition of animal manures and sewage, which form the more important organic fertilizers used, has also been described. The composition of animal manures is highly variable, and is dependent on the animal species, nature of its food, handling and storage of the manure, and the climatic conditions.

Organic matter generally varies between 15% and 34%, total nitrogen 0.3% to 1.7%, and phosphates (P_2O_5) 1.25% to 2.96% (FAO, 1977*b*). Animal manures applied in a pond in a fine particulate or colloidal state stimulate heterotrophic growth of bacteria by providing the necessary surface area for their attachment and facilitating their mineralization. The mineral fraction is also directly used for photosynthesis, but because of restricted light penetration, organic manures seem to promote the production of zooplanktonic organisms. Heterotrophic production is more rapid at the soil–water interface and this gives rise to abundant benthic growth, which is of special importance in pond farms raising benthic feeding species.

Though organic fertilizers are generally preferred in aquaculture, many farms use inorganic fertilizers, either alone or in combination with organics. Inorganic fertilizers are in simple inorganic compound form containing at least one of the primary nutrients, namely nitrogen, phosphorus and potassium (NPK). They may also include nutrients such as calcium, magnesium and sulphur, and trace elements such as copper, zinc, boron, manganese, iron and molybdenum.

While NPK fertilizers seem to be favoured in North America, West European aquaculturists lay greater stress on phosphate fertilizers. The main disadvantage of phosphate fertilizers is that phosphorus compounds are not easily soluble in water and are absorbed by the bottom soil or mud and often converted into insoluble compounds. These are released only when micro-organisms change them into assimilable forms. If the phosphate fertilizer is mixed with organic manure, the inorganic compounds are gradually converted into organic phosphate compounds in manures, and these could more readily be utilized for plankton production.

In theory, fertilizers are applied in quantities that can be readily utilized by the ponds for production of the desired food organisms. Dosage has to be adjusted according to the intensity of grazing, avoiding the development of algal blooms, which may create unfavourable environmental conditions. Account is also taken of the nutrient status of the water and soil in the ponds before the application of fertilizers. It can readily be seen that under farm conditions it is far from easy to determine the precise dose of fertilizers to be applied. Over-fertilization can happen, therefore, but since even ponds that exchange water occasionally stop outflow for a certain time, the chances of the fertilizers or the biota produced finding their way to the open waters are minimal. Most farms normally discharge water only at the time of harvest. In tidal ponds where there is more frequent exchange of water, benthic algal complexes that become dislodged from the bottom and float are collected and dried, to be used as feed or fertilizer. If they are not removed regularly, they may be wafted to the pond banks by winds and waves, where they settle and decompose. Wastes that accumulate on the pond bed are either removed to be used as fertilizer for terrestrial plants or dried by exposure to the sun after dewatering, and allowed to mineralize after appropriate liming and controlled inundation. Liming is also adopted to neutralize acidity of pond soils and to sterilize sediments. Though large quantities of lime may have to be used on acid sulphate soils, the environmental impacts appear to be minimal.

7.6   Residues of Biocides and Biostats

In land-based farms, especially pond farms, chemicals and other toxic substances are commonly used to control predators, pests and weeds, usually as part of pond preparation before stocking with larvae, fry or fingerlings. Teaseed cake, derris powder and rotenone are widely used to clear aquaculture waters of pests and predators. Teaseed cake is the residue of the seeds of *Camellia drupisera* after extraction of the oil, and generally contains 10–15% saponin. A dose of 216 kg teaseed cake, together with 144 kg of quicklime per hectare is applied on the pond bottom after reducing the water level. The toxicity disappears within 2 to 3 days. Rotenone, which is another plant product, is widely used for the same purpose as derris powder. It contains 4–8% of the active ingredient rotenone ($C_{23}H_{22}O_6$) when applied at levels of 0.5 ppm, and the toxicity is reported to disappear within about 48 hours (Hall, 1949). Even at concentrations of up to 20 ppm, under tropical conditions the toxicity ceases after 8 to 12 days (Alikunhi, 1957).

Several other non-selective toxins, including DDT, endrin, aldrin and 2,4-D are used for eradicating pests and predators, but repeated washing of the pond is carried out after treatment to remove the toxic effect. This results in the discharge of at least small quantities of the toxin into the receiving waters. To avoid this, the use of the agricultural weedicide sodium pentachlorophenate (PCP–Na) is recommended to eradicate predatory and weed-fish, because it decomposes when exposed to direct sunlight and the toxicity is reduced by 90% after about three hours.

Selective toxins that are used in pond farms include Bayluscide (5,2-dichloro-4-nitro-salicylicaniline-ethanolamine) and nicotine (commercial tobacco dust) to control snails and polychaete worms, and technical benzene hexachloride (BHC) (containing 6.5% gamma isomer) and the insecticide 'Sevin' to control crabs infesting pond dikes.

Infestation of emergent, floating and submerged weeds is a persistent problem in pond farms, especially in tropical areas. Because of the ease of their use and the relatively rapid results, various types of herbicides were used in eradicating them. In recent years, aquaculturists have become more selective in the use of herbicides, as a result of the recognition of the dangers of using some of them. Among the herbicides that continue to be used extensively are 2,4-D (2,4-dichlorophenoxyacetate) and Diquat (6,7-dihydrodipyrido (1,2-a:2′,1′-C) pyrazidiinium salt) for

Table 7.5 Use of chemicals in European fish farms. (Quantities in kg except formalin and hyamine which are in l. Numbers of farms involved are shown in parentheses.)

Chemical		Average quantity per farm	per t fish	Average concentration (mg l⁻¹)
Malachite green	(12)	15 (8)	0.2 (8)	1.1 (5)
Formalin	(10)	138 (6)	0.7 (6)	101.0 (3)
Chloramine	(4)	28 (3)	–	–
Diuron	(3)	60 (2)	4.0 (2)	–
Copper sulphate	(2)	50 (1)	–	2.0 (1)
Calcium hydroxide	(2)	300 (1)	–	–
Hyamine	(2)	2	0.14	0.35
Sodium chloride	(1)	200	0.025	–
Dipterex	(1)	3	2.0	–

(From: Alabaster, 1982)

controlling floating and emergent weeds by foliage application. Copper sulphate ($CuSO_4$) and simazine (2-chloro-4,6 bis(ethylamine)-truazine) are used for the eradication of submerged weeds, even though they are non-selective and have a long-term effect on pond productivity. Copper sulphate is the most widely used and economical weedicide, especially for controlling algal growths. Copper sulphate pentahydrate (CSP) is reported to be very effective against several algae. Anhydrous ammonia has been used successfully to eradicate dense growths of submerged weeds, and its residual effect is as a nutrient in the water.

Even though the persistence of the toxicity of all the chemicals used for controlling weeds, pests and predators in tropical pond farms is not precisely known, the culture practices make it unlikely that they have any major impact on the environment. Since waste discharges from the farms are only occasional and long after the application of chemicals, it is considered unlikely they will have any residual effect, even though their decomposition rates and possible absorption of breakdown products by pond sediments have not been determined. Studies of the insecticide Dipterex (2,2,2-trichloro-1-hydroxyethyl phosphate), used for controlling ectoparasites, seem to indicate that it is rapidly hydrolysed under tropical pond conditions (Haque & Barua, 1988).

Table 3.1 in Chapter 3 presents data on the extent of chemical usage in fish farms in the UK for disinfection, control of predators or treatment of diseases. Table 7.5, from Alabaster (1982), summarizes the available information on the use of chemicals in European fish farms. The data indicate that the amount used in relation to fish production is relatively small and the concentrations used are low.

The environmental impact of using chemicals is more direct in aquaculture in open waters, such as mollusc farming in the coastal areas and cage and pen farming of finfish. A number of chemicals are used in oyster culture to control pests and predators. Lime, trichloroethylene and dichlorobenzene, and the pesticide Sevin, are commonly used. Chemicals are seldom applied on fish in cages, mainly because of obvious problems in effective administration. In special circumstances, as in the case of salmon net-pen farming in Norway, when pesticides such as Neguvon and Nuvon are used to control fish lice, there have been lethal effects on crustaceans in the vicinity of the farm (Egidius & Moster, 1987). Even though the information available now indicates only localized effects of the use of chemicals, there is an obvious lack of adequate critical field data. The present conclusions are based on dilution in the environment and persistence and water solubility of the chemicals (Beveridge, 1984; Ackefors & Södergren, 1985; Weston, 1986a). Pesticidal chemicals are among the most difficult to study because of their great biological potency, even in ultra-micro quantities. They are often beyond chemical detection limits and therefore have to be studied by a combination of bioassay and chemical techniques (Walker, 1961).

Rosenthal *et al.* (1988) have drawn attention to the potential effect of chemicals introduced through construction materials, such as plastic additives and antifouling compounds. Several types of additives are often used in the manufacture of plastics, including stabilizers, pigments, antioxidants, UV absorbers, flame retardants, fungicides and disinfectants (Zitko, 1986; Rosenthal *et al.*, 1988). Many of these are essentially toxic to aquatic life, but the extent of their effect under open-water conditions outside the farms has yet to be determined in view of their low water solubility, slow rate of leaching and dilution.

An antifoulant that has been studied from the point of view of its effect on the environment is tributyl tin (TBT). Compounds of TBT are widely used on pleasure boats, in harbour construction and on marine cages, because of high toxicity to fouling organisms, low toxicity to man and the long duration of antifouling protection. The toxic effect of TBT was first observed in France (Arcachon Bay) on oyster beds in areas frequented by pleasure boats. Poor spat production, larval abnormalities and shell malformation were observed, and TBT was implicated as the causative agent (Alzieu & Portmann, 1984; Alzieu & Héral, 1984). Subsequent work has confirmed the toxic effect on other forms of marine life as well, and shown that it causes reproductive failure or growth abnormalities in molluscs (Maguire, 1986; Cardwell & Sheldon, 1986; Paul & Davies, 1986; Stang & Seligman, 1986; Thain, 1986). Recent studies show that TBT can cause mortality in cage-cultured fish, and accumulate in their tissue (Short & Thrower, 1986a and b). The accumulated data on adverse environmental effects, including the effect on open-water aquaculture, have led to restrictions on the use of TBT in a number of European countries and in Canada, as well as the initiation of a re-evaluation in the USA.

Very few countries appear to regulate and monitor the use of chemicals in aquaculture, partly due to the absence of appropriate information on their environmental effect. Table 7.6 gives a list of chemicals used in aquaculture. Besides pesticides and weedicides used in aquaculture

Table 7.6 Chemicals used in aquaculture (modified from Barg, 1992).

	Use
Therapeutants	
Acetic acid	Ectoparasites
Formalin	Ectoparasites
Malachite green	Ectoparasites and fungus
Acriflavin (or Porflavin hemisulphate)	Ectoparasites, fungus and bacteria
Nuvan (Dichlorvos)	Salmon lice (Ectoparasitic copepods)
Common salt or brine	Ectoparasites
Buffered iodine	Bactericide
Oxytetracycline	Bactericide
Romet 30 (Sulphadimethoine and orthoprim)	Bactericide
Tribrissen (Trimethoprim-sulphadiazine)	Bactericide
Hayamine 3500	Surfactant–bactericide
Benzalkonium	Bactericide
Chloride	Bactericide
Chloramine T	Bactericide
Vaccines	
	Vibrio anguillarum
	Aeromonas salmonicida vibrio
	Enteric red-mouth disease
Anaesthetics	
MS222 (tricaine methane sulphonate)	
Benzocaine (ethyl-paraminobenzoate)	
Carbon dioxide	
Disinfectants	
Calcium hypochlorite	
Liquid iodophore, e.g. FAM 30	
Sodium hydroxide	
Water Treatment	
Lime	
Potassium permanganate	
Copper sulphate	

and the chemicals that inadvertently affect aquaculture and the environment, there are several drugs widely used as therapeutic agents. While there is hardly any restriction on the use of therapeutic agents in a large majority of countries, in some it is extremely difficult to get the necessary approval from the authorities concerned.

In his study of fish farm effluents in the UK, Solbe (1982) found that malachite green and formalin, along with Buffodine, which is a combination of the two, accounted for more than half the list of chemicals used. It should be noted that malachite green is not approved for use in the USA because it is considered to be carcinogenic.

Table 7.7 Maximum concentrations* of chemicals reported to be discharged per year into receiving waters in the UK.

Chemical	Number of farms	Mean (mg l^{-1})
Malachite green	25	0.61 ± 0.70
Formalin	17	15.2 ± 48.1
Hyamine	4	0.55 ± 0.97

* Arithmetic mean ± standard deviation. (From: Solbe, 1982)

As the use of chemicals to control epizootics is mainly in land-based rearing facilities where the outflow of water can be controlled, the concentration of chemicals in the effluent discharged into the receiving waters is relatively low. The maximum concentrations of three chemicals discharged from fish farms in the UK, reported by Solbe (1982), are given in Table 7.7. Considering the total amount used per year of 176 kg of malachite green, 1230 litres of formalin and 77 litres of hyamine, the maximum concentration observed in the effluents does not appear to be very significant.

The use of antibiotics in aquaculture has grown significantly, even though its value in controlling diseases has become controversial. Antibiotics are used to cure diseases that have occurred in hatcheries or production farms. They are normally administered through medicated feeds which are available commercially. Because of its broad spectrum of activity, chloramphenicol is favoured in mollusc, shrimp and prawn hatcheries, but many other antibiotics have also been tried in different concentrations.

Rosenthal *et al.* (1988) cite the Norwegian marine salmon culture industry's use of antibiotics in 1984, to be 6223 kg oxytetracycline, 7820 kg Tribrissen R, 5500 kg nitrofurazolidone and 9 kg sulphamerazine. According to later reports, the use of antibiotics has gone up to 18 000 kg for producing 120 000 tons of salmon in Norway in 1989. It has been estimated that only 20–30% of the antibiotic administered orally is actually taken up by the fish. The rest reaches the environment and is bound to particles and sediments (Bjorklund *et al.*, 1990).

Samuelson (1989) found that the degradation of oxytetracycline in seawater is rapid and the half-life is 128 and 168 h at 5 and 15°C, respectively, under 24 h illumination, and 390 and 234 h, respectively, in darkness. When the antibiotic layer on the sediment is covered with new sediment the degradation becomes slower. The major concern caused by indiscriminate use of antibiotics is not so much the concentration in the effluents, but its undesirable effects and its real effectiveness in preventing infections in aquaculture. Widespread use of oxytetracycline in south-east Asian shrimp ponds is reported to have resulted in the development of resistant strains of pathogenic vibrios, which has caused major problems in the treatment of vibrio infections.

Bacteria are known to develop resistance to antibiotics rapidly. Furthermore, a species of bacteria can transmit its resistance to an antibiotic to a different bacterial species by genes contained on extra chromosomal pieces of DNA called plasmids (Brown, 1989). It is believed that bacteria will develop resistance to most, and possibly all, antibiotics with which they are challenged. The selection of antibiotic-resistant strains of bacteria occurs during chemotherapy.

Austin (1985) identified an increased proportion of resistant bacteria in the effluent when using oxolinic acid, potentiated sulphonamide and oxytetracycline, even though the resistance was short-lived and returned to pre-treatment levels within a period of about nine days. He drew attention to the ineffectiveness of antibiotics to prevent or even cure diseases of cultured stock, and the risk of transference of antibiotic resistance to normal bacteria within the human gut if numbers of antibiotic-resistant bacteria are ingested. Human pathogens do survive in both fresh and marine waters and so there is a danger to human health. There is, therefore, a strong case for discouraging the use of antibiotics in aquaculture. The use of antibiotics is being phased out in large-scale aquaculture.

It is known that chemotherapeutic products used in the treatment of bacterial, fungal and parasitic diseases remain in fish for varying periods and can become a public-health hazard. The elimination process includes metabolism and excretion of drugs and it has been demonstrated that elimination half-lives are longer at lower temperatures. Additionally, there appears to be a tendency for high concentrations to remain for longer periods in certain tissues, such as the kidney, liver, skin, scales, bones and uveal tract. Drug residues have been detected in blood and/or muscle tissues from experimentally treated fish from a few days up to more than 100 days after administration (Schmid, 1980). However, because of differences between experimental and farming conditions and the differences of drugs used in previous studies, it is difficult to draw definite conclusions.

A number of countries have adopted regulations regarding withdrawal periods for drugs administered to fish, but only a few have taken into account the influence of temperature and other variable factors. Pending the acquisition of adequate pharmacokinetic data on drugs, Rasmussen (1988) has suggested the adoption of general withdrawal periods of 80 days after the final application in waters with temperatures below 10°C, and 40 days after final application in waters with temperatures above 10°C, after the use of chemicals and/or therapeutics for the prevention or treatment of fish diseases.

Steroids, including specific hormones, are increasingly used in commercial fish culture for inducing maturation and spawning (mammalian gonadotropic hormones, human chorionic gonadotropin (HCG), fish pituitary homogenates and follicle-stimulating hormone (FSH)), and for sex reversal (methyltestosterone, ethyltestosterone, estrone, ethynylestradiol and stilbestrol). Studies so far seem to indicate that at the dosage of hormones used, there is no potential human health hazard. Johnstone *et al.* (1983) investigated the elimination of methyltestosterone from fish treated for sex reversal and confirmed this conclusion. In commercial farms, only the brood stock are sex-reversed and the marketed fish are not treated. This ensures further safety. Incorporation of hormones in the feed has been tried experimentally to enhance growth rates of fish, but it has not yet found application in commercial production.

7.7   Algal Blooms

Enrichment of an aquaculture farm from the decomposition of waste feed, faeces, metabolic wastes and, in many cases, added fertilizers, has already been referred to in Chapter 3. The hypernutrification caused by this can result in algal blooms in the farm itself or in the vicinity of the farms (especially in the case of cage farms in open waters and pen culture in eutrophic lakes). This may involve changes in the species composition of the phytoplankton and a reduction of the organisms that form the food of cultivated species. The growth of autotrophic phytoplankton and its uptake of dissolved nitrogen and dissolved organic phosphorus can have the beneficial effect of controlling the nutrient budget in fish ponds and consequently the nutrient status of their effluents.

In experimental studies in a gilthead seabream (*Sparus aurata*) pond having a water retention time of 2 days, where the major nutrient pathways consisted of fish excretion and phytoplankton intake, Krom & Neori (1989) found that the ammonia-N concentration in the outflow was reduced from 100 μm to undetectable levels, and dissolved organic nitrogen to nearly half its maximum value, when phytoplankton communities built up after a plankton bloom.

Oxygen consumption by dense algal communities during the night may result in low levels of oxygen concentration in the absence of photosynthesis to renew it. In addition, the senescence and anaerobic microbial degradation of large masses of algae can give rise to low levels of dissolved oxygen and promote the growth of undesirable macrophytes.

Another possible effect of algal blooms is the formation of algal toxins, which may accumulate in cultured organisms, especially molluscs. Although there have been mortalities of farmed fish (Jones *et al.*, 1982) and shellfish, which were apparently due to algal blooms caused by emissions from aquaculture farms, the major blooms of dinoflagellates or red tides that have seriously affected aquaculture do not appear to be connected with farm wastes. For example, the catastrophic blooms of the little-known algal species *Chrysochromulina polylepis*, which wiped out entire salmon farms and threatened the salmon cage industry in Norway, originated in Sweden and then moved northwards along the west coast of Norway. Though the exact cause of this algal bloom is not known, it is believed that unusual climatic conditions and ocean currents provided the appropriate conditions for its occurrence.

Blooms of *Gyrodinium aureolum*, which caused the mortality of marine organisms in north European waters, have been reported by Tangen (1977), Jones *et al.* (1982) and Doyle *et al.* (1984) in the UK and Ireland, but there is no apparent relationship with the discharge of aquaculture farm wastes. Similarly, the red tide caused by the dino-flagellate *Pyrodinium bahamense* var. *compressium*, which has caused paralytic shellfish poisoning (PSP) in many tropical countries around the world, is not related to aquaculture.

The amnesic shellfish poisoning (ASP) caused by eating cultivated blue mussels from Prince Edward Island in Canada (Bird & Wright, 1989) has been traced to a massive bloom of the pennate diatom *Nitzchia pungens* forma *multiseries* that produced the neurotoxin domaic acid. Here again, the unusual weather conditions and the run-off of nutrients after a sudden and severe rainstorm are suspected to be the cause, even though the addition of nutrients from the mussel beds has not been ruled out. In all known cases, aquaculture, especially of molluscs, has been adversely affected by algal blooms.

7.8 Bacterial Communities

Considerable attention has been devoted in aquaculture research to patho-genic bacteria, but very little to natural microbial communities in the farm itself and in the environment affected by aquaculture. Limited studies and observations on the effect of antibiotic use on bacterial populations and the potential dangers of transference of antibiotic resistance between bacteria have been discussed in Section 7.6.

The enrichment of bottom sediments in land-based farms, as well as in cage and pen culture in open water, provide favourable conditions for the multiplication of bacterial populations and the enhancement of specific bacterial communities. The fate of coliform bacteria in fish farms receiving sewage has already been referred to in Chapter 6. Available information on the effect of environmental conditions of an aquaculture farm on coliform bacteria is not conclusive. In enclosed water areas, as in ponds, a drastic reduction of coliforms has been observed (Carpenter *et al.*, 1974). Bergheim & Selmer-Olsen (1978) reported no change in faecal coliform populations in the receiving waters of effluents from a large freshwater trout farm in Norway. A 12-months' monitoring of a rainbow trout farm in a Scottish freshwater loch by Smith *et al.* (quoted in University of Stirling, 1988) did not find any coliforms in the water column throughout the year. Similarly, Korzeniewski & Korzeniewska (1982) failed to find any enhanced growth of coliforms in a Polish fresh-water lake adjacent to rainbow trout cages. In contrast, Haavisto (1974) reported elevated concentrations of coliform and streptococci bacteria in effluents from freshwater trout farms in Finland, and Hinshaw (1973) in hatchery effluents from the USA.

Niemi & Taipalinen (1980) showed that counts of indicator species of bacteria (total coliform, faecal coli and faecal streptococci) were high in effluents of fish farms in Finland, even though there was no evidence of disease risk to man or livestock (Sumari, 1982). It is known, however, that several forms of bacteria can colonize the intestines of fish and provide suitable conditions for their multiplication. The 1980 study in Finland detected indicator bacteria in the receiving waters, but observed their number in the farm effluent to be very low.

Investigations by Allen *et al.* (1983), Austin (1985) and Austin & Allen-Austin (1985) have shown that there is very little effect on the bacterial populations as a result of trout farming and, in most cases, the bacterial numbers in the effluents were similar to or less than the numbers in the influent waters. No change was noticed in the composition of the bacterial populations either.

Though there is no adequate evidence, the possibility of marine culture being a net contributor of bacteria in the receiving water was indicated in a study of a coastal turbot-rearing facility. A substantially higher concentration of bacteria and changes in the bacterial composition, compared to those in the inflow water, were observed in the effluents. It is very likely that this particular situation is related to site-specific farm management problems.

The possibility of disease transmission from cultured fish from pond farms through effluents, or directly from cages and pens in open waters, has often been raised. Chapter 9 will discuss cases of inadvertent introductions of pathogenic bacteria into new areas, resulting in infection of wild fish or shellfish, but there is no evidence of aquaculture leading to the increased incidence of diseases caused by pathogenic bacteria already present in the region. Moriarty (1986; 1997) advocates probiotics to displace the pathogens responsible for the occurrence of shrimp diseases in farms by competitive processes and the growth of inhibitors. Sonnenholzner and Boyd (2000) question this on the basis of experiments on the use of commercially available probiotics.

There are instances of disease being transmitted through rivers and other waterways to cultured stocks in pond farms. A recent example is the epidemic of mortality caused by an ulcerative syndrome that has occurred in many countries of Asia. The exact causative agent is still not known, but the disease has spread through regular inflows or inundation of farms during floods.

Chapter 8
Pattern and Effect of Waste Discharges

Chapter 7 reviewed the sources and processes of waste production. In considering the environmental effect of aquaculture practices, the quantity and frequency of waste discharges from the farms into their environment are important factors. Although in most cases the discharges go into water bodies, there are cases where farms are dewatered along with the wastes by pumping or other means on to land areas, for example, the case of 'undrainable' ponds in South-Asian aquaculture. As mentioned in Section 3.4, the quantity and quality of the wastes are affected by the density of cultured stocks, the type and quantity of foods and feeds provided (Ackefors, 1999) and the retention time of water in the rearing facilities. Suspended solids and dissolved nutrients, especially nitrogen and phosphorus, have been identified as the most important waste products affecting the quality of the receiving waters and their environment. As will be discussed below, the pattern of discharges of these wastes greatly influences their effect.

8.1   The Nature of Waste Discharges

The mode and frequency of discharges obviously depend on the culture system adopted and the relevant water management procedures. Based on this, two main discharge systems can be identified: (1) seasonal or minimum discharges, and (2) continuous discharges. The continuous discharges can be directly from the cultured animals into the environment, as from cages, or through special drainage systems, as in pond farms. The production rate of wastes in farms varies considerably according to the season and the age composition of stocks under culture, especially because of changes in feed consumption rates. This is likely to be reflected in the quantity of wastes in the farm discharges.

Similarly, some of the routine farm procedures can affect the quality and quantity of wastes discharged. For example, cleaning of tanks, raceways and other rearing facilities can result in sudden increases of wastes in the discharges.

Harvesting or other disturbances in ponds and pens can result in resuspension of solids, which may get carried into the receiving waters.

In his studies of a number of channel catfish ponds in Auburn (USA), Boyd (1978) found that the settleable matter concentration ranged from 0 to 1.2 ml l^{-1} during the draining phase of harvesting (when 95% of the water was discharged), as against the instantaneous maximum of 3.3 ml l^{-1} suggested by the Environmental Protection Agency (EPA-USA, 1974*b*), but the water discharged during seining to remove the fish contained >3.3 ml l^{-1} settleable matter (range 0.5–100 ml l^{-1}). Concentrations of other pollutants, except nitrate, also increased during seining.

The use of special harvesting devices in mollusc farms can have a noticeable impact on benthic communities. Peterson *et al.* (1987) have described the ecological consequences of mechanical harvesting ('clam kicking') of the hard clam *Mercenaria mercenaria* in shallow estuarine habitats in North Carolina (USA). Intensive clam kicking resulted in seagrass biomass declining by 65% below expected levels. Recovery started only 2 years later, and remained comparatively low even 4 years later. Even light clam kicking resulted in a decline of 25%, but full recovery occurred within a year. Density of bay scallops (*Argopecten irradians*) in seagrass beds showed significant decline. The density or species composition of small benthic macro-invertebrates were not affected adversely, probably as a result of rapid recolonization. The use of mechanical oyster harvesting is also suspected to affect intertidal benthic invertebrate communities associated with oysters, even though reliable evidence is lacking.

The University of Stirling report (1988) draws attention to other sources of aquaculture-related wastes discharged into coastal areas. Net cages which are not coated with antifouling paints are removed and cleaned at frequent intervals. After allowing the nets to dry for some time, high-pressure water jets are often used to dislodge encrustations (see Fig. 8.1). Sometimes, chemicals such as copper sulphate, iodophor disinfectants or biological washing powders are reported to be used. When the cleaning is done on shore, there is a likelihood of some of the organic matter and chemicals being discharged into the sea. The damage caused to shore life due to this, however, is not considered very significant.

As mentioned in Section 3.3, a substantial part of present world aquaculture production comes from pond farms, which are dewatered only at the time of harvesting. Water levels may be lowered to facilitate harvesting, and after harvest complete dewatering is carried out to dry the ponds and prepare for the next crop. Application of fertilizers is stopped several days before harvesting, and feeding ceases at least a day before harvest. Suspended solids and dissolved nutrients are therefore likely to be minimal in the effluents. The ponds usually have a thick deposit of detritus which is either removed to be used as fertilizer for terrestrial crops or dried on the pond bed and treated to serve as fertilizer when the ponds are refilled for the next cycle of culture. Although there is very little documented information on the environmental effect of waste discharges from such pond farms, it would appear that they do not contribute in any significant way to hypernutrification or eutrophication.

Figure 8.1 Net cages being cleaned with high-pressure water jets to dislodge encrustations.

In a comparison of aquaculture systems, Folke and Kautsky (1989) observe that salmonid farming is about 15 times as harmful to the sea-bottom as mussel rearing, since under a 40-t marine fish farm, sedimentation increases twenty-fold, while under a 100-t mussel culture it increases only three-fold. The net effect of a mussel cultivation cycle is the removal of nutrients from the environment, as no food is added. Kaspar *et al.* (1985) consider mussel culture to be a means of counteracting eutrophication, as large-scale mussel culture can lead to nutrient decrease, even though the pattern of nutrient cycling is altered. In fact, polyculture of bivalves is recommended as a general means of sustainable aquaculture (see below).

Seasonal rain-fed fish ponds that dry up during the dry season do not discharge wastes into water bodies, but may contribute to ammonia in the atmosphere as a result of evaporation of nitrogen-rich pond water. Recirculation systems generally fall into this category also as waste discharges are minimal.

Continuous or periodic discharges of wastes occur from freshwater and coastal pond farms and raceways that normally maintain a continual flow of water of constant or variable velocities. The load of wastes in the outflow would naturally depend on the culture practices, but a major advantage of these systems is that the discharges can be controlled to a certain extent, and that special waste treatment facilities can also be incorporated. By adjusting the flow of discharges and the flushing time, the adverse effects of discharges on the environment can be greatly reduced.

Cage and pen culture facilities and mollusc farms emit wastes into the environment on a more or less continuous basis. Though the level of discharges can be comparatively small in relation to time, high stock density in the rearing units can contribute to substantial accumulations if the water exchange in the environment is not adequate for rapid dispersal and decomposition of wastes (Barg, 1992).

8.2 Polyculture

Polyculture is one of the means of developing aquaculture as a sustainable activity (Grant, 1996; Neori *et al.*, 1996; Boyd & Heaseman, 1998). Bivalve aquaculture is well developed in many countries where molluscs form a favoured menu in human diets. Since no external feeding is required, they constitute a primary product, and serve as biofilters in an integrated system. Bivalves feed on existing plankton and enhance water quality through feeding. They can be managed as a constituent of polyculture with marine and brackish water plants that make use of dissolved nutrients (Negroni, 2000). No net addition of nitrogen to the environment is involved as a result of feeding with external food items.

There are some potential constraints, such as food limitations and carrying capacity of he environment (Grant & Bacher, 1998; Grant, 1999), competition by fouling organisms (Clareboudt *et al.*, 1994) and potential changes in the ecosystem as it reaches bivalve dominance. The production of faeces and pseudofaeces by suspension feeders is often alleged to be a source of organic loading, but there is no net addition of organic matter to the environment since bivalves feed on material already present. Carrying capacity, which is particularly significant, can be assessed by simulation modelling, despite problems of transferring site-specific models to new locations referred to in Chapter 3.

Poor site characterization leads to reduced reliability of model production. Permanent, extensive culturing may bring about changes in benthic communities (Tenore *et al.*, 1982). In spite of all these constraints, polyculture remains an effective means of achieving sustainability (Grant, 1996; Hunt *et al.*, 1995; Boyd & Heaseman, 1998; Grant & Bacher, 1998). It contributes to the dual role of maintaining the environment and increasing the ability to produce protein (Bodvin *et al.*, 1996). For example, polyculture as an integral part of shrimp culture can manage effluents in shrimp ponds. However, Canzonier (1998) points out that bivalves concentrate and accumulate pathogenic micro-organisms and chemical substances in polluted waters. The tradition of consuming raw or lightly cooked molluscs carries the risk of affecting public health. Several measures of depuration are followed in industrialized countries (see Fig. 12.1).

Chapter 9
Introduction of Exotics and Escape of Farmed Species

9.1 Species Diversity

There are several reasons for the introduction and transplantation of species and selected strains from one area to another in aquaculture, as in agriculture and animal husbandry. Though there are a large number of culturable species, there are only a few that can be considered domesticated and for which culture technologies have advanced to a level that gives an aquaculturist some assurance of success. Furthermore, the locally available species may not have the same or similar culture characteristics such as temperature tolerance, feeding, growth and production, or consumer acceptance. The local species may not be suitable for developing a recreational fishery, as required for social and economic reasons. Species depleted in an area due to diseases or over-fishing may have to be replaced or rehabilitated.

Deliberate introductions and transplantations of several species have, therefore, been carried out around the world, e.g. the brown and rainbow trout, salmon, the common carp, tilapia and oysters. Introductions of aquatic animals and plants have been carried out not only for aquaculture but also for other purposes, such as establishing sport and commercial fisheries, for biological control or for ornamental aquaria. There have also been accidental introductions, such as the classic example of the original introduction of *Tilapia mossambica* in Java, Indonesia. Other types of accidental introduction that have occurred are of pathogens that the introduced species may have been carrying, or of organisms that came in unnoticed with the water used for transport of the introduced species.

Though comparative ecological data before and after introduction are not available in all cases to determine whether the introductions have been harmful or not, there are some that have been harmful to the environment and others that are considered, overall, to be beneficial. As pointed out by Wilson (1965) and Munro (1986), it is not normally possible to predict the consequences of introductions, because the ecological background of colonization by introduced species is not fully understood. Munro (1986) cites two examples of the unpredictability of colonization, namely the failure of anadromous salmonids to colonize various parts of the world where they have been introduced, and the explosion of the carp population a century after its introduction in Australia.

Transplantations and stocking of semi-enclosed or open waters, as well as ranching of anadromous fishes, have been undertaken in different parts of the world to enhance existing stocks or to rehabilitate stocks that were depleted due to various causes. As a result of technical improvement and appropriate scaling-up of operations, many of these programmes have eventually become successful (Jhingran & Natarajan, 1979; McNeil, 1979). The on-going organized attempt to develop 'culture-based' fisheries in Japan is another instance of enhancement of existing stocks.

9.2 Ecological Effects of Introductions

The most important ecological effect of an introduction or transplantation of a species to a new environment is its influence on the local plant and animal life. Even though the introduction may be only for culture in protected farms, escapes into the environment, accidentally or otherwise, are considered inevitable. Some species may find the conditions unsuitable for building up a sizeable spawning population, and may, therefore, not become a dominant species. Others may, under the same conditions, reduce or eliminate constituents of the local fauna through competition or predation. There are records of the reduction or disappearance of species from natural water bodies as a result of the introduction of exotic fish.

The introduction of the brown trout, *Salmo trutta*, and the rainbow trout, *Oncorhynchus mykiss*, in natural bodies of water has been reported to have caused the disappearance of several indigenous species. It is believed that the disappearance of the freshwater mullet (*Trachsjiostoma euronotus*), kurpers (*Sandelia capensis*), *Amphyluis hargesi*, *Galaxias* spp. (Munro, 1986) and *Oreodaemon gnathlambae* in southern Africa, and *Trichomicterus* spp. in Columbia (Welcomme, 1988) are reported to be due to trout introduction.

Rainbow trout introduction is believed to be responsible for the elimination of some native cyprinidonts (*Orestias* spp.) in Lake Titicaca in Peru (Villwock, 1963; Everett, 1973), some *Galaxias* spp. in certain areas of Australia (Cadwallader, 1978; Jackson, 1981), a decline of gallaxiids in New Zealand (McDowall, 1968), the extinction of the New Zealand grayling, *Prototractes oxyrhynchus* (Allen, 1949), and the reductions in the native crayfish *Paranephrop* sp., a freshwater crab and a frog in New Zealand (Fish, 1966).

The introduction of *Tilapia mossambica* in Lake Butu in the Philippines is reported to have resulted in the near elimination of *Mistichthys luzonensis* stocks (Baluyut, 1983). *Cichla ocellaris* introduction is reported to have caused damage to stocks of native fishes in Lake Gatun,

Panama, and *Basilichthys bonariensis* is also implicated in the decline of *Orestias* spp. in Lake Titicaca.

It should be stressed, however, that the effect of ecological changes brought about by other concurrent causes have not always been adequately studied and eliminated to lead to the definite conclusion that the adverse effects have been due to the introductions. This is not to suggest that the introduced species would not have had an effect on the indigenous fauna, but only to indicate that it is not always possible to isolate the effects of introductions from other environmental perturbations from natural or man-made causes.

With enhanced interest in prawn and shrimp farming, there is an increasing number of international exchanges of these groups. The giant freshwater prawn (*Macrobrachium rosenbergii*) has been introduced from South-east Asia into several countries in tropical, sub-tropical and even temperate climates in almost every continent. The Japanese shrimp *Penaeus japonicus* has been introduced into many countries in Asia, southern Europe, west Africa, southern United States, and Central and South America. The widely cultivated Indo-West Pacific shrimp *P. monodon* has been introduced into some of the Central and South American countries for aquaculture.

As though in exchange, some of the east Pacific species of penaeids (*P. stylirostris*, *P. vannamei*) have been imported into east Asia (Philippines and Taiwan). Taiwan has also imported *P. brasiliensis* and *P. schmitti* from South America. It is suspected that these introductions have caused the spread of viral diseases among cultivated shrimps.

A crustacean introduction that has caused some controversy is that of the Louisiana red crayfish *Procambarus clarkii* in Japan and in southeast Africa. It was introduced into Japan as a food supply for bullfrogs. Penn (1954) reported that it colonized and became a pest, eating rice crops and undermining rice field dikes. A similar experience is reported from Uganda where it also became a pest. Obviously, these introductions were of a casual nature, without appropriate follow-up measures for controlled production, as the same species is reported to be successfully cultured in Spain (Sandifer, 1986). The signal crayfish *Pacifastacus leniusculus* was introduced into Scandinavian countries in 1960 to replace the European noble crayfish *Austropotamobius pallipes*, which were decimated by the 'crayfish plague' caused by the fungus *Aphanomyces astaci*. As *P. leniusculus* is immune to the disease, it has established itself in Scandinavian lakes, which provide favourable environmental conditions for its growth. No adverse effects of its introduction have been reported.

9.3 Transmission of Diseases

Next to salmonids, carp and tilapia, the species that have been very widely transplanted or introduced are the oysters. Though concerns were expressed earlier about the possible effects of exotic species on local oysters, so far there appears to be no evidence of direct competition or incompatibility. Nevertheless, there are several instances of associated disease outbreaks and incidental introductions of pests and weeds.

Massive introduction, spread over a decade, of the Pacific oyster *Crassostrea gigas* from Japan and British Columbia to the French coast, to replace the declining populations of the so-called Portuguese oyster *C. angulata* is one of the sustained, and in some respects successful, cases of marine introductions of exotics. However, a series of severe epizootics of *C. angulata*, and of the European flat oyster *Ostrea edulis*, occurred during this period, more or less simultaneously. Also at the same time, the protozoan parasite *Marteilia refrigens* affected the oyster-growing areas, closely followed by epizootic infections by another protozoan parasite, *Bonamia ostreae*. The two diseases also caused serious mortalities among *O. edulis* populations.

These simultaneous occurrences gave rise to the suspicion that the introduction of *C. gigas* may have played a role in the disease outbreaks, though extensive research has not yet yielded any evidence of a direct relationship. There is, however, evidence that the seaweed *Sargassum muticum*, introduced inadvertently with the Pacific oyster, has spread along the coasts of France and England, in some places forming dense beds and obstructing other uses of the coastal areas. The seaweed *Undaria pinnatifida* has also been introduced like this, along with the oyster.

Other incidental introductions reported are of the parasitic copepod *Mytilicola intestinalis*, which infects the gut of shellfish, the oyster drill *Urosalpinx cinerea*, and the American slipper limpet *Crepidula fornicata*.

Several reports exist of diseases spread by fish introductions. The grass carp *Ctenopharyngodon idella*, which has been introduced from the Far East to Europe, has been reported to have transmitted the cestode *Bothriocephalus acheilognathi* (Ivasik *et al.*, 1969). The extinction of *Orestias* spp. from Lake Titicaca after the introduction of rainbow trout is ascribed partly to the passive introduction of sporozoan parasites.

Furunculosis, a systemic bacterial disease caused by the obligate fish pathogen *Aeromonas salmonicida*, was probably introduced into the UK from Denmark with brown trout, and spread through movements of farmed trout (University of Stirling *et al.*, 1990). Furunculosis has apparently been introduced into Norway with infected salmon smolts from Scotland, and this has caused severe losses to the Norwegian salmon farming industry.

The best example of the introduction of a pathogenic organism in relation to salmon farming is of the monogenean *Gyrodactylus salaris* into Norway, where it caused massive mortalities of salmon populations, and in some areas their total eradication. It was introduced into Norway with smolts from Sweden and gradually spread to several hatcheries.

Besides direct transmission of a disease, introductions can also have catalytic effects (Mills, 1982). It has been suggested that appropriate defences against diseases of the native species may be lost with the introduction of new species or strains of fish. This can result in the disease organisms becoming more dominant in the environment, with selection of more virulent strains that eventually infect the native species. Mills (1982) indicates that this was the case with the occurrence of viral haemorrhagic septicemia (VHS) in brown trout in the UK after the introduction of rainbow trout from North America, which were very susceptible to the disease.

The catastrophic losses of the native crayfish *Austropotamobius pallipes* due to the 'crayfish plague' in English rivers is ascribed to the fungal agent *Aphanomyces astaci*, introduced with the American signal crayfish (Alderman *et al.*, 1984). Sindermann (1986*b*) draws attention to the risk of spreading shrimp virus diseases as a result of extensive transfers and introduction of shrimp species.

Infectious hyperdermal and haematopoietic necrosis (IHHNV) disease has been observed in Hawaiian shrimp culture facilities rearing *Penaeus stylirostris*, introduced from Panama. Juveniles and adults of *P. vannamei* were also recognized as carriers of the virus. Though the possibility of native populations being infected by the introduced pathogen has caused concern, there is, as yet, no evidence of it having occurred.

The risk of transmission of pathogens is not limited to the introduction of eggs and the young of cultivated species, but also to the import of live animals and ungutted carcasses for sale to consumers. The danger is greater in filter-feeding molluscs, which are commonly imported alive for sale. The practice of holding imported live animals for long periods before sale to consumers adds to the risk of transmission of pathogens, as well as of pests, as this will involve the changing of water in the containers and discharge of the waste water into the environment.

Meyers (1984) has cited evidence that shellfish can accumulate viruses and bacteria at much higher concentrations in their tissue than in the surrounding water. When grown in polluted environments containing human pathogens, there is the likelihood of their accumulating the pathogens, and acting as carriers if moved into new environments. Several virulent viral pathogens of finfish have been isolated from bivalve molluscs. These include: 13 p_2 reovirus, causing chronic granulomatous hepatitis and occasional pancreatitis in rainbow trout; JOV-1, causing mortality in young trout; salmonid infectious pancreatic necrosis (IPN) virus; molluscabirna viruses, causing mortality in trout fry; chum salmon virus (CSV); infectious haematopoietic necrosis (IHN) virus; and piscibirnavirus (Meyers, 1984). Depuration of the bivalves in clean or sterilized water removes most of the pathogens harmful to humans, but many fish viruses survive and are virulent even after 60 days of depuration in clean water (Meyers, 1980). It is interesting to note, however, that in spite of experimental observations, there appear to be no known instances of fish disease having been transmitted in this manner.

9.4   Control of Introductions

As indicated at the beginning of this chapter, the need for the introduction of exotic species and improved strains exists and will continue. Irrespective of whether some of the observed ill-effects are due to the introductions or not, it is only logical to assume that the presence of an exotic species in an environment will contribute to changes occurring due to other causes as well. The fact that no evidence of adverse effects has been found would not be a valid reason for complacency, especially since very few of the introductions have been preceded by adequate scientific enquiry. So there is an obvious need to prevent indiscriminate introductions. A number of governments have prohibited the introduction of exotic species without the permission of appropriate authorities. Such prohibitions have been extended in some countries to include live fish and shellfish, as well as ungutted carcasses (as in the case of salmon), to prevent transmission of pathogenic organisms.

To avoid indiscriminate introductions, Turner (1949) suggested the following criteria be adopted in selecting species for introduction. The species should:

- fill a need, because of the absence of a similar desirable species in the locality of transplantation;
- not compete with valuable native species to the extent of contributing to their decline;
- not cross with native species and produce undesirable hybrids;
- not be accompanied by pests, parasites or diseases that might attack native species;
- live and reproduce in equilibrium with its new environment.

These basic criteria are still valid and the American Fisheries Society (Anon., 1973), the International Council for the Exploration of the Sea (ICES, 1972 and 1979), and Kohler & Stanley (1984) have strengthened the arguments for critical evaluation and proposed sets of protocols for implementation of these criteria. Based on these protocols, Kohler & Stanley (1984) have suggested a detailed review and decision model for evaluating proposed introductions of aquatic organisms.

The American Fisheries Society protocol consists of the following steps.

(1) *Rationale.* Reasons for seeking an import should be clearly stated and demonstrated. It should be clearly noted which qualities are sought that would make the import more desirable than native forms.
(2) *Search.* Within the qualifications set forth under Rationale, a search of possible contenders should be made, with a list prepared of those that appear most likely to succeed, and the favourable and unfavourable aspects of each species noted.

(3) *Preliminary assessment of the impact.* This should go beyond the area of Rationale to consider the impact on target aquatic ecosystems, including the effects on game and food fishes and waterfowl, on aquatic plants and on public health. The published information on the species should be reviewed. A preliminary study of the species in its biotope should be made.

(4) *Publicity and review.* The subject should be open for review and expert advice should be sought. At this point, thoroughness is required. No import is so urgent that it should not be subject to careful evaluation.

(5) *Experimental research.* If a prospective import passes the first 4 steps, a research programme should be initiated by an appropriate agency or organization to test the import in confined waters, e.g. experimental ponds, etc. This agency or organization should not have the authority to approve its own results or to effect the release of stocks, but should submit its report and recommendations for evaluation.

(6) *Evaluation or recommendation.* Again publicity is in order. Complete reports should be circulated among interested scientists and presented for publication in the Transactions of the American Fisheries Society.

(7) *Introduction.* With favourable evaluation, the release should be effected and monitored; the results should be published and circulated.

The ICES Code of Practice, first adopted in 1973, revised in 1979 and approved by the countries bordering the North Atlantic, relates to the introduction of marine species, defined as 'any aquatic species that does not spend its entire life cycle in freshwater'. The revised code of practice is reproduced below.

Revised code of practice to reduce the risks of adverse effects arising from the introduction of marine species

At its Statutory Meeting in 1973, the International Council for the Exploration of the Sea adopted a *Code of Practice to Reduce the Risks of Adverse Effects Arising from Introduction of Non-indigenous Marine Species.* At its Statutory Meeting in 1979, the Council adopted a revised code as follows:

1. Recommended procedure for species prior to reaching a decision regarding new introductions. (This does not apply to introductions or transfers which are part of current commercial practice.)

    a. Member countries contemplating any new introduction should be requested to present to the Council at an early stage information on the species, stage in the life cycle, area of origin, proposed place of introduction and objectives, with such information on its habitat, epifauna, associated organisms, potential competition to species in the new environment, etc., as is available. The Council should then consider the possible outcome of the introduction, and offer advice on the acceptability of the choice.

b. Appropriate authorities of the importing country should examine each 'candidate for admission' in its natural environment, to assess the justification for the introduction, its relationship with other members of the ecosystem, and the role played by parasites and diseases.

c. The probable effects of an introduction into the new area should be assessed carefully, including examination of the effects of any previous introductions of this or similar species in other areas.

d. Results of b. and c. should be communicated to the Council for evaluation and comment.

2. If the decision is taken to proceed with the introduction, the following action is recommended:

a. A brood stock should be established in an approved quarantine situation. The first generation progeny of the introduced species can be transplanted to the natural environment if no disease or parasites become evident, but not the original import. The quarantine period will be used to provide opportunity for observation for disease and parasites. In the case of fish, brood stock should be developed from stocks imported as eggs or juveniles, to allow sufficient time for observations in quarantine.

b. All effluents from hatcheries or establishments used for quarantine purposes should be sterilized in an approved manner (which should include the killing of all living organisms present in the effluents).

c. A continuing study should be made of the introduced species in its new environment, and progress reports submitted to the International Council for the Exploration of the Sea.

3. Regulatory agencies of all member countries are encouraged to use the strongest possible measures to prevent unauthorized or unapproved introductions.

4. Recommended procedure for introductions or transfers which are part of current commercial practice:

a. Periodic inspection (including microscopic examination) by the receiving country of material prior to mass transportation to confirm freedom from introducible pests and disease. If inspection reveals any undesirable development, importation must be immediately discontinued. Findings and remedial actions should be reported to the International Council for the Exploration of the Sea.

b.   Inspection and control of each consignment on arrival.

c.   Quarantining or disinfection where appropriate.

d.   Establishment of brood stock certified free of specified pathogens.

It is appreciated that countries will have different attitudes to the selection of the place of inspection and control of the consignment, either in the 'country of origin' or in the 'country of receipt'.

(Also see *www.ices.dk/.*)

Though the ICES Code of Practice represents an international uniform policy, it is considered somewhat idealistic and difficult to impose (Sindermann, 1986*b*) and quite stringent (University of Stirling, 1988). Mann (1979) pointed out that by restricting itself to the containment of adverse biological effects of introductions, the code has failed to address the economic, sociological and political pressures that may exist to support introductions. Though approved by the governments concerned, application of the code is reported to be poor, 'with little progress except as a consequence or reaction to crises' (Sindermann, 1986*b*). However, the stringent nature of the proposed procedure, though not strengthened by legislative measures, can be expected at least to deter inessential transfers.

While it is relatively less difficult to determine whether the imported species brings in parasites or diseases, prediction of ecological effects based on controlled experiments has many limitations. It is important that monitoring the effects of introductions should be carried out on a long-term basis in order to adopt possible measures as soon as any signs of environmental deterioration are noticed.

9.5   Genetic Dilution due to Escape of Farmed Animals

The escape of native or exotic species from culture installations happens in spite of possible precautions, as a result of accidents and natural disasters such as flooding, storms and typhoons. Caged fish stocks are particularly susceptible to losses, because of damage to cages by predators, vandalism and poaching. The species may also be deliberately released into the environment, as in the case of salmon parr released into river systems to enhance homing stocks. Accidental losses generally involve only small numbers, except in the case of natural disasters or when a cage farm has been badly damaged. The general observation is that hatchery-reared individuals have a lower survival rate and ability to compete in open waters than wild ones. This is possibly due to their greater susceptibility to predation and lower ability for inter-specific competition.

Even though there are no precise estimates of the survival of the escapees, there are records of well-established rainbow trout fisheries at trout cage culture sites in Scotland. The annual harvest of 5 t from Loch Fad (in Scotland), which is about 2.5% of the total cage production, is reported to be based on escaped trout (Phillips *et al.*, 1985). Phillips *et al.* (1985) also present data on catch rates of rainbow trout in Loch Charn to show that the rate has increased after the establishment of cage culture. It is reported that 1–4 t of escaped rainbow trout has been harvested from a Polish lake, about 5% of its total cage production in 1977 (Penczak *et al.*, 1982).

It is important to note that these observations are from lakes and lochs rather than from more exposed areas. Investigations seem to show that the escapees feed near the cages, presumably on spilled feed and the enhanced growth of food organisms in the surrounding enriched waters. Increased commercial catches of rainbow trout have also normally been in the neighbourhood of cages. The physical structure of the cages, the presence of caged fish, and the abundance of benthos and zooplankton are possible causes for the congregation of fish around cages (Phillips *et al.*, 1985, Kilambi *et al.*, 1976).

The extent of competition of the escapees with local species and stocks depends on their feeding habits and the ecological niches they occupy. However, from the environmental point of view, the more important concern is probably of the escapees interbreeding with wild fish and altering the genetic make-up of wild populations, particularly by introducing genes that are less suited for local conditions.

Most of the widely cultivated fish and shellfish stocks, such as salmon, trout, carp, tilapia, oysters, mussels and shrimps, belong to several distinct stocks. The distinctions are further enhanced by selection processes that are aimed at producing animals that perform well under culture conditions. With successive farm-raised generations, the difference in genetic characters between cultured and wild stocks becomes much more pronounced.

Ståhl (1983) has described the differences in genetic make-up of wild and cultured Atlantic salmon. Strong spatial genetic heterogeneity was generally observed within and between stocks from different river systems, whereas hatchery stocks show a significantly lower genetic variability. A reduced variation within hatchery stocks and a lower genetic divergence between them were observed.

Though available evidence is not very conclusive, indications are that hatchery-reared fish are less fit for survival in the wild than progeny of wild fish. If this is so, it will result in the introduction of inferior genes into the populations' gene pools. If the farm stock establishes breeding stocks in natural water bodies, or succeeds in interbreeding with wild stocks, as has been observed in some cultivated species, there is a real danger of loss of genetic diversity.

The possible harmful effects of inbred and polyploid fish and shell-fish on natural populations of these species by competition and inter-breeding, or even of replacing them, have been pointed out by Thorgaard and Allen (1988). It has been suggested that if introductions are required, it will be preferable to use sterile organisms as they are likely to have the least negative impacts on natural populations. Even though sterile hybrids or triploids might in some cases interfere with reproduction of natural stocks in behavioural or other non-genetic ways, their genetic impact may be only minimal. Fertile hybrids should help in introdu-cing beneficial genes or chromosome segments into domesticated stocks, but can be recommended for use only in closed systems with little risk of their escape into natural environments.

As in many other aspects of the environmental impact of aquaculture practices, assessments of risks of release of genetic material into aquatic ecosystems are based on inadequate scientific data, including those on net social benefits. There is obviously a need for rational consistent programmes to assess the risks involved and to structure a meaningful dialogue with the public on the information derived (Gregory, 1988).

9.6   Guidelines for Management of Movement of Live Aquatic Animals

As mentioned in Section 9.4, efforts to control introductions started several decades ago, but hitherto they have not been widely enforced for various reasons. The stringent nature of the procedures proposed and the lack of legislative support have been advanced as the major reasons for their low enforcement. It has now been recognized that control measures have to be flexible as the predicted ecological effects of introductions are not always fully substantiated by field observations. Guidelines have to be followed, not only nationally but also regionally and, if necessary, internationally, to be effective (FAO/NACA, 2000). It is also now recognized that introduction and transfer of live aquatic animals may be unavoidable as an essential part of farming and stock enhancement. Live animal trade is increasing with advances in trans-port within and between countries. Many sport fisheries are sustained by the import or export of fertilized ova, fry or fingerlings. Seedlings of bivalves are often transported over long distances for farming or stock enhancement. There is therefore considerable controversy over protec-tion of native biodiversity and fortuitous spread of pathogens. Because of this and the limited value of earlier efforts, the Fish Disease Com-mission of the Office International des Epizootics (OIE) developed recommendations and protocols for preventing the international spread of animal disease as a part of its *International Aquatic Animal Health*

Code (*www.oie.int/eng/normes/fcode*). These relate to health surveillance of animals involved in domestic and international trade. Guidance is offered for actions required to reduce the risks associated with the introduction and transfer of species.

Knowledge of the health status of aquatic animal populations is essential for risk assessments of pathogen transfer. Health certification and quarantine measures are major components of the proposed health management system. It is believed that some degree of risk is inherent in the introduction and transfer of aquatic animals, and so all health management procedures should aim at being practical and cost-effective, and capable of implementation within the available administrative structure. Health management of aquatic animals should include capacity building for competence to supervise and implement regulatory measures. Facilities to hold live aquatic animals for disease inspections at the importing border will be required to diagnose the cause of specific diseases. Quarantine regulations have to be developed to prevent the transfer of disease agents with live aquatic animal movements at exporting and importing points.

The main aim of health management of movements of live aquatic animals is to define the zones of disease occurrence and report any possible spread of important pathogens. Management of zones of diseases requires an appropriate level of diagnostic and monitoring ability, and reporting facilities. These should be accompanied by adequate regulatory control mechanisms, where responsibilities are clearly assigned to designated competent authorities. Regional or international reporting authorities can alert importers to ensure appropriate certification and quarantine action.

Chapter 10
Pathogens in the Aquatic Environment

All forms of aquaculture are susceptible to outbreaks of disease, as many pathogenic bacteria are normal inhabitants of the aquatic environment. Both in aquaculture facilities and in external aquatic environments, the occurrence of disease is a complex interaction between the host species, disease agents and the environment. In farm environments, outbreaks of disease are greatly influenced by the susceptibility of the hosts (Falsted *et al.*, 1993), the virulence of the pathogens and adverse environmental conditions. Farming practices may favour disease occurrence, as in the case of intensive and semi-intensive systems of production characterized by high stocking densities, increased stress of stocks, intensive feeding and inadequate water exchange. As a result, it is not uncommon for outbreaks of epizootics to occur in aquaculture farms. Aquaculture literature describes a large number of infectious and non-infectious diseases caused by various pathogenic agents.

On the other hand, disease outbreaks are less common in open water environments, even though the pathogens and host species may be present. The host species may live healthy normal lives in the continuous presence of pathogens, and only when environmental stresses occur will the balance change, favouring the dominance of the pathogen. If the host species are not able to adjust to the changes, disease and mass mortality occur.

In the context of the environmental impact of aquaculture, this chapter is restricted to an examination of possible increases of pathogenic populations in open environments, and disease transmission to wild stocks as a result of aquaculture practices. Even though pathogens may be physical, chemical or biological, the main focus of attention here will be the biological agents that cause infectious diseases, and the environmental changes that give rise to disease conditions.

10.1 Occurrence of Pathogens

Pathogens are always present in the environment. The main concern, therefore, is of introducing new disease organisms or causing environmental deterioration that can result in increased populations and virulence of indigenous pathogens, and disease outbreaks in wild stocks as a result of waste discharges from aquaculture farms, or the introduction or transplantation of culture stocks. Distinction has to be made between indigenous opportunistic pathogens that can cause disease only when the host's resistance is lowered or when unusual circumstances favour its growth and development, and obligate disease-causing organisms that cannot survive in nature unless susceptible or carrier hosts are present.

Introduction of exotic species or the transplantation of indigenous species often results in significant increases in the diversity of pathogenic organisms in the environment. Many instances of fortuitous introductions of disease organisms through introduction or transplantation of exotic species for aquaculture have been cited in Chapter 9. In most of the documented cases, the pathogenic organisms were transferred to the aquaculture farms, and the main threat of disease outbreaks has been to the farmed species. Undoubtedly, pathogens can be transferred from the farm to the external environment and the wild stocks, but documented evidence seems to show very low prevalence of the pathogens and lack of clinical symptoms in infected wild stocks, as in the case of infectious pancreatic necrosis (IPN) virus and enteric red-mouth disease (ERM) (Munro *et al.*, 1976; Phillips *et al.*, 1985).

In studies of a freshwater reservoir fishery in England, Allen *et al.* (1983) found no major imbalance in the composition of bacterial flora leaving the reservoir fishery via the effluent. The incidence of disease in farms can be explained by the higher density of cultured animals than in natural habitats and differences in water quality conditions.

In shellfish culture in open waters, the risk of infecting wild populations appears to be marginal, as in most cases shellfish are cultivated in areas where natural populations did not exist previously, or have been decimated. As pointed out by Sindermann (1986a), pathogenic roles are assumed by certain organisms in cultivated populations of fish and shellfish, such as monogenic trematodes and some protozoans that are often rare and innocuous parasites in natural populations. Virulent pathogens are usually associated with epizootics in natural populations, whereas facultative pathogens tend to emerge as causes of epizootics in cultured populations. There are many cases of farms being infected by pathogens brought in with water inflows from open environments, leading to mass mortalities under unfavourable environmental conditions.

Despite the absence of conclusive evidence of major infections of wild stocks from aquaculture farms, it is necessary to take precautionary measures, as very little research has been done to define the role of aquaculture in the outbreak of diseases in wild stocks. There is a definite possibility of infections being carried from one farm to the other through inflows from a common source. For example, pond farms situated along a common watercourse, which serve as both the source of water inflows and also drainage of waste water, can easily become infected through discharges from an infected farm upstream. Similarly, neighbouring cage and shellfish farms can easily exchange infections depending on nearness and water-flow patterns. The disastrous collapse of Taiwanese shrimp farming is at least partly ascribed to the use of contaminated drained water from adjacent ponds.

Dumping carcasses of infected fish into natural waterways is another practice that contributes to increased populations of pathogens in the environment. Animals that die from infectious diseases are a major source of pathogens, as clinical illness is associated with significant increases in the pathogenic populations. It is important, therefore, to remove sick and dead individuals to minimize the presence of pathogens in the environment. Collection and destruction of dead fish in such a way as to prevent dissemination of pathogens is therefore a legal requirement in fish farms in many areas. Similarly, when a farm is infected with viral diseases for which there are no control measures, farmers are required to destroy the stock to prevent the spread of the disease to other farms and to the external environment. This is a practice more commonly followed in the control of communicable diseases in animal husbandry, for which farmers require to be compensated.

10.2   Environmental Causes of Disease

The environmental impacts of aquaculture, such as eutrophication, can create conditions favourable for disease outbreaks. Bacterial populations in natural waters are related to the trophic state. Grimaldi *et al.* (1973) found pathogenic fungal infections to be common in alpine lakes, and Korzeniewski & Korseniewska (1982) found higher bacterial counts in the vicinity of rainbow trout cages in Lake Letowo in Poland, evidently as a result of environmental changes brought about by waste discharges from cages. Despite these observations, there is very little evidence of epizootics occurring in natural stocks as a result of aquaculture practices. However, the possibility of even local environmental degradation enhancing the susceptibility of wild stocks to infections has to be considered until it can be proved otherwise, and all possible measures taken to prevent the degradation of the environment, even if it is only to protect aquaculture farms in the neighbourhood.

The risk of discharging pharmaceutical compounds used in farms for controlling diseases into the aquatic environment was discussed in Chapter 7. Increasing use of anti-microbial compounds poses the risk of release of the bioactive components into receiving waters. Uptake of these with potable water supplies can lead to allergic reactions in humans. However, it has been observed that the use of oral medication leads to only negligible quantities of anti-microbial compounds being discharged through effluents from aquaculture facilities. Though the use of bath treatments can cause higher concentrations, all available evidence indicates that the discharge of anti-microbial compounds does not constitute a real environmental problem.

Experimental work has shown that degradation of the antibiotic oxytetracycline in seawater occurs within 128 and 168 hours at 5 and 15°C, respectively, under continual illumination (see page 69), although in darkness the time increases to 190 and 234 hours, respectively. It has also been observed that most oxytetracycline gets bound to particulates and deposits at the bottom of the farm. In conditions where sedimentation is slow, the half-life of oxytetracycline has been reported to be 32 days. Biodegradation may be retarded by deposition of sediments over the antibiotic, and then the half-life may be doubled to 64 days (Samuelson, 1989). Antibiotic-containing sediments can affect fauna in the environment. The possible development of antibiotic-resistant microbial communities, as a result of antibiotic use in aquaculture farms, was discussed in Chapter 8.

10.3 Controlling the Spread of Communicable Diseases

As discussed in Chapters 3 and 9, some aquaculture practices have the potential to cause the spread of diseases, especially dangerous communicable diseases, among both farm stocks and stocks in the external environment. Controlling this requires co-operative efforts both nationally and internationally. Farmers, traders, veterinarians, fishery administrators and legislators have to recognize the dangers involved.

In earlier years, the small size of the aquaculture industry was used as a justification for inaction in some countries. The scenario has changed in recent times and aquaculture is now recognized as a growth industry, contributing substantially to food production and to national economies in the majority of countries.

The time is ripe for effective action based on adequate knowledge of the etiology of diseases and their transmission. As described in Section 9.6, central to communicable disease control is an integrated health-management programme. Thompson *et al.* (1973) describe preventive measures required for fish disease control as follows:

– improvement in design, operation and inspection of fish culture establishments;
– inspection, quarantine and certification of transferred, traded or imported stocks;
– development of a disease registry, reporting and information service;
– adoption and enforcement of adequate and comprehensive legislation.

Though the general principles involved are the same, some of the above measures are to be implemented nationally and others internationally. While harmonized national legislation or an international convention is an essential element in controlling the import of disease agents, an effective inspection programme of farms and hatcheries is probably the most efficient means of preventing interfarm transfers and release to the environment of disease agents. Periodic inspections have to be made mandatory through appropriate legislation, and provision has to be made for the destruction of infected stocks, when required, under appropriate conditions.

Some countries of Europe and North America have established periodic inspections and adopted guidelines or codes of practice for controlling the import of exotics. Several countries, including developing countries, have enacted regulations to control the import of live fish and eggs to prevent the spread of diseases. In most countries, some form of certification of freedom from disease is required for clearance of consignments for import.

Chapter 11
Birds and Mammals in Aquaculture

Aquaculture becomes closely associated with several species of birds and mammals because of the location of farms on or near their natural habitats. Pond farms are often built on marshes that are important fly-ways, nesting and over-wintering areas for many species of birds. Some mammals are also characteristic of wetlands, particularly musk-rats, mink and otters. Many of them are predatory in habit, and farms with dense populations of fish and shellfish offer convenient foraging grounds. So the aquafarmer considers them a threat to his stock and has to take all possible measures to prevent predation. From an environmental point of view, the farm and the farmer's activities, especially his efforts to protect farm stocks, may constitute a threat to these species of birds and mammals (NCC, 1989). Some of them may be protected by wildlife preservation acts in certain countries and eradicating them by shooting or other means may not be permissible without the special permission of competent authorities. This involves a clear case of conflicting interests, requiring adequate understanding of the nature and extent of the damage caused to determine appropriate measures of reconciliation.

Several methods of controlling bird predation are practised by farmers, with varying degrees of success, anti-predator nets being the most commonly used protective device. While small pond and race-way farms can be covered with nets of suitable mesh size, it is not easy to cover extensive pond farms in this manner. Devices such as flash guns, sirens, klaxon horns, gongs, scarecrows, bamboo rattles, bells and windmills fitted with mirrors have all been tried with initial success, but in the course of time the birds learn to ignore them.

Among mammalian predators, the most destructive are probably the otters (*Lutra* and *Aeonyx*). They attack relatively large fish, eating the best parts and leaving the rest. Farmers try to prevent their entry into farms by suitable fencing, and use traps to catch them. The grey seal (*Halichoerus grypus*) and the common seal (*Phoca vitulina*), and the mink (*Mustela vison*), are reported to be important predators of caged fish in Scotland. As relatively rare species, the seals receive some protection by law in this country.

The main impacts of aquaculture on bird life are:

- physical damage by protective devices and deliberate killing;
- disturbance by aquaculture activities and scaring devices;
- disruption of natural habitats.

The impacts on mammals are very similar to those listed above for birds, and include:

- deliberate killing and live trapping by farmers;
- accidental entrapment in anti-predator nets and fencing;
- disruption of natural habits by the establishment and operation of farms.

11.1   Effect of Birds on Aquaculture Farms

Birds are major predators or pests in aquaculture farms and the losses sustained are very significant (Pillay, 1990). A pelican can consume between 1 and 3 t of fish per year. According to du Plessis (1957), ten breeding pairs of cormorants in captivity will catch about 4.5 t of fish per year. Herons may cause losses up to 30–40% of fry and juvenile fish in a pond farm. A heron may consume as much as 100 kg of fish per year. Bird predation in shrimp ponds is reported to decrease production by about 75% in Texas (USA).

Birds are attracted to farms, especially pond farms, as they are often located in marsh and mangrove areas with large colonies of birds and mammalian predators. The shallow waters of coastal ponds provide ideal conditions for birds to prey on dense stocks of cultured species. Cormorants, fish eagles, herons and kingfishers are considered the most destructive predators. Large flocks of cormorants can drive fish into shallow areas of ponds by flapping their wings, and then prey on them. Herons and cormorants are also the main predators in freshwater fish farms in Scotland (University of Stirling *et al.*, 1990). The majority of fish eaten by these birds are probably less than 20 cm in length, though cormorants will eat larger fish than herons.

Differences in feeding efficiency have been noticed between age classes of fish-eating birds (Morrison *et al.*, 1978; Quinney & Smith, 1980). It has also been observed that all age classes of these birds have a higher intake of food in fish farms than in their natural habitats. Diving bird predators seem to be attracted to feed near cage farms because of the increased concentration of wild fish near fish-rearing cages.

There is also some seasonality in the abundance of bird predators, related to behaviour patterns or weather conditions. For example, in temperate areas they tend to congregate in larger numbers during winter or in autumn, when juvenile birds become independent (Meyer, 1981).

Grebes, shore birds (Charadriiformes), gulls (Family Laridae) and crows are mainly competitors for food and are seldom predators. In cage farms they feed on spilled fish feed and from beneath hoppers. The droppings of large flocks of crows and gulls in surrounding waters and on floating structures of off-shore cage farms, where they roost at night, can pose public health hazards because the droppings contain pathogenic organisms. There are a number of parasites that could be transmitted by birds. Herons appear to be the final host for fish tapeworms (cestodes), and herons, gulls, grebes and goosanders are the final hosts for fish flukes (trematodes), which infect both wild and farmed fish in their larval stages.

Birds can act as mechanical carriers of viruses, causing infectious pancreatic necrosis (IPN) and viral haemorrhagic septicaemia (VHS), as the viruses have been isolated from regurgitated food several hours after feeding on infected fish. The possibility also exists of infection being transmitted through faecal matter, as the IPN virus could be isolated from the faeces of herons several days after ingestion. Peters & Neukirch (1986) have shown that it is possible to reinfect trout fry from this faecal matter. Though there are many other references to possible transmission of pathogenic organisms, such as salmonellae, faecal coliforms and the bacterium *Edwardsiella tarda*, no clear linkage between fish diseases and contamination of water bodies by bird faeces has been demonstrated. It is quite likely, however, that predator damage on fish can make them susceptible to a number of diseases.

11.2 Effect of Aquaculture on Birds

Activities in an aquaculture farm, such as feeding, fertilizer application, fishing, grading, maintenance work and changing of nets in cage farms, can cause disturbance to birds. Increased communications, including road and boat traffic, is generally associated with the establishment of farms, and the extent of disturbance depends on the size and nature of operations. Both cage culture, which involves constant care and the use of powered boats for transporting humans and materials to the farm, and large land-based pond farms that employ mechanical equipment for feeding, water management and aeration, cause more disturbance than some other forms of aquaculture, such as shellfish farming in tidal areas. It has been suggested that such disturbances may cause birds to leave the farm area, either temporarily or permanently. This does not appear to have been substantiated, even though some species may be scared initially by such disturbance. Predatory birds attracted to farms by the easy availability of feed appear to become accustomed to the noise and other forms of disturbance, including scaring devices. The localized nature of most such disturbances has not yet been observed to have had any detrimental effect on species such as herons and cormorants (University of Stirling *et al.*, 1990).

Disruption due to the reduction and alteration of the natural habitat of birds is another possible effect of the establishment of aquaculture farms. Though there is no clear-cut evidence, it is likely that the presence of a farm may make an area unattractive to certain species of wildfowl, divers, grebes, etc. Also, the attraction of opportunistic species of birds to the farms by the availability of feeds may bring about greater competition for local bird populations.

The most important impact on bird life can be caused by deliberate killing and by entrapment in anti-predatory devices. Where birds are not protected by legislative measures, farmers often resort to shooting to kill or to scare them from the site. Even in areas where the shooting of birds is prohibited, special permission can be obtained by a farmer to protect his stock. For example, in Scotland, where it is an offence to kill, injure or take almost any wild bird, the competent authority is empowered to issue licences to kill or take wild birds in order to prevent 'serious damage' to 'fisheries'. Such permits are normally given to freshwater fish farms, though not to marine farms.

Despite this, the killings usually involve only a small proportion of the bird populations, as shown by studies of grey herons. There is, as yet, no evidence that the population magnitude of herons has been adversely affected in any noticeable manner by such shooting (Meyer, 1981; van Vessem *et al.*, 1985). Shooting does not deter fish-eating birds from visiting fish farms. Being highly mobile they continue to congregate in farms where food is freely available. Reported observations indicate that most of the birds killed at farms belong to the first year-class, and the level of killing does not appear to affect population size. Some farmers may use poisoned bait and traps to capture birds, but the mortality caused by these methods does not appear to be significant.

Anti-predator nets and other protective devices are reported to cause more damage and mortality than shooting. Cages are often protected by some type of anti-predator net. Ross (1988) and the University of Stirling (1988) reviewed the methods of using predator netting in Scottish cage farms. Top nets fixed on the top edges and stretched across the cage are commonly used to exclude surface-feeding birds, including herons. This is meant to prevent birds from taking fish directly from the water surface within the cage by, for example, gulls, terns, cormorants, shags and gannets. Underwater nets are suspended below the cage, outside the net-bag holding the fish, protecting the four sides and the bottom.

Some farms use weighted sheets of netting suspended from the cage frames to protect only the four sides. These are meant to exclude diving birds such as shags and cormorants, which attack fish through the side of the cage nets, as well as to prevent attack by marine mammals. Tank, race-way and small pond farms are generally protected by nets or a series of strings stretched across at a suitable height. Chains of polystyrene floats laid along pond margins or steepened pond sides are also reported to be useful in deterring herons.

There are confirmed reports of deaths of fish-eating birds such as cormorants, shags and guillemots which have become trapped in underwater anti-predator nets. Birds may also get entangled or trapped in top nets. Despite the fact that many birds die or are injured by entangling or entrapment in anti-predator nets, the magnitude of effect on their population size is not considered high, except in the case of cormorants, where there is some indication that mortality may be higher than in other species (University of Stirling, 1988).

11.3   Effect of Aquaculture on Predatory Mammals

The nature of the effects of aquaculture on mammals appears to be very similar to the effects on birds described in Section 11.2, but there is comparatively less quantified information on the affected populations. Otters are considered the most destructive mammalian predators in most forms of aquaculture, in both tropical and temperate climates. Otters live in the immediate vicinity of water and burrow into the banks under the roots of trees. They are nocturnal in habit and hunt for fish mainly on clear nights. It has been observed that in cage farms, otters climb into the cages from walkways to eat fish, or chew holes in the cage nets to remove fish. They have been reported to kill many more fish than they can eat, the loss being as much as 80% of the stock.

The University of Stirling (1988) reports that seal and mink predation is more serious in cage farms in the UK. Like otters, mink also appear to kill and damage more fish than they can eat. Seals attack caged fish on moonlit nights from underneath the cage, biting the fish through the net. The methods adopted for controlling predation by mammals are killing and trapping, provision of anti-predator nets and fencing. Shooting is the most common means of killing, but trapping is more common for controlling otters. Traps, including live traps, are also used in many areas. Mink are caught in baited live-trap cages and left to drown (Ross, 1988). Studies have shown that mink populations have a high turnover and that the whole population can be replaced within about three years. This suggests that shooting or trapping is unlikely to affect population size in the long term. Underwater anti-predation netting is used in cage farms to protect fish from mammalian predators. It is reported that some farmers set anti-predator nets in such a way as to trap seals rather than to protect the fish stocks.

The effects of aquaculture on mammals have not been sufficiently investigated to arrive at reliable conclusions. In some areas, such as Scotland, otters are legally protected and it is an offence to cause deliberate disturbance in their place of shelter, but there is provision for licensing the taking or killing of otters to prevent serious damage to livestock or fisheries. Original assumptions that increasing disturbance may be partly responsible for the otters' declining population are now considered to be exaggerated in areas of sufficient cover, where their tolerance level has been shown to be high (Jefferies *et al.*, 1984). It has been shown that they are often not even scared by human activities and evidently get used to farm conditions. In fact, some believe that the growth of fish farming has only benefited otter populations.

Shooting of seals is regulated in Scotland to conserve stocks, and a permit is required to shoot them during closed seasons. Licences are issued only to prevent significant damage to fisheries. It is believed that entanglement in anti-predator nets in fish-farm areas accounts for almost an equal number of deaths as by shooting.

Chapter 12
Safety of Aquaculture Products

The safety of products for consumption is of prime concern from the point of view of managing aquaculture, as well as ensuring public health. Official regulatory bodies in many countries specify maximum permissible concentrations of toxic substances or the number of harmful bacteria that a product may contain, in order to ensure that unfit or unwholesome food does not reach the consumer. Even though not usually covered by regulations, unattractive appearance and tainting of products affect their marketability.

Water quality and culture practices play important roles in determining product quality. Of major concern in aquaculture are environmental contaminants, effects of using waste water in farms, pathogenic contamination of shellfish and toxic algal blooms. There is also growing concern for the potentially adverse effects of genetically modified organisms (GMO).

12.1 Breeding Programmes and Genetically Modified Food Products

Selective breeding of cultured fish has been undertaken in order to maximise production and to enhance some of the characteristics that help consumer acceptance, as in the case of Atlantic salmon (Gjedren, 1985), coho salmon (Hershberger *et al.*, 1990) and tilapia (Eknath, 1993). The technology needed for genetic modifications, which is now used extensively in agriculture and animal husbandry, has not yet been developed for aquaculture, but if or when this happens, the consequences will be very far-reaching. Selective breeding has already gained acceptance in raising species like tilapia, which was earlier considered a low-quality fish with limited markets. This has given the hope that many low-quality freshwater fish can find remunerative markets if subjected to genetic manipulation.

Losses due to diseases in marine and coastal aquaculture are sure to trigger interest in genetic modification for the acquisition of immunity to pathogenic organisms. The present practice of chemical control has proved to be environmentally harmful and not sufficiently effective.

One has to take into account the possibility of genetically modified products proving toxic to humans. Transferring single genes to produce immunity may result in the escape of the transplanted genes to wild species. There is also the potential risk of cultivated species being fed on feeds containing ingredients that are genetically modified.

12.2   Environmental Contaminants

Both in enclosed farms as well as in open-water aquaculture facilities, farmed species can be exposed to contaminants, some of which are persistent and have the capacity for bioaccumulation. As aquaculture is sited in inland or inshore waters, there is a greater chance of contamination through domestic, agricultural and industrial wastes.

Tainting of products by environmental contaminants has received special attention in the fisheries sector because of its significance in consumer acceptance of products. The gills, skin and gut of fish can rapidly absorb odoriferous chemicals from water and cause tainting of the flesh. Tainting compounds may be formed in the flesh by metabolism of innocuous precursors absorbed from water or by direct ingestion and absorption (Howgate & Hume, 1986).

The most common type of taint encountered in aquaculture is the muddy, earthy flavour of fish grown in fresh or brackish water environments, caused by geosmin, 2-methyl-isoborneol, mucidone or similar compounds. It is reported to be produced by blooms of species of actinomycetes and blue-green algae, especially in semi-stagnant water systems. Though such off-flavours do not pose any public health problem, they can result in economic loss to the farmer. Tainting of fish and shellfish by industrial chemicals and by mineral oil have also been reported (Brunies, 1971; Shumway & Palensky, 1973).

Controlling the occurrence of off-flavours in culture systems, which involves controlling the growth of micro-organisms that synthesize the causative compounds in the enriched medium, is an extremely difficult process. When tainting has occurred, depuration in clean water is the most practical way of removing it. Lovell (1979) reports that off-flavours produced by actinomycetes and blue-green algae in pond-cultured catfish can be removed by depuration in 5 to 15 days, depending on water temperature and the intensity of the off-flavour. Tainting from fuel or crude oil or industrial chemicals requires a much longer time of depuration, of a month or more.

12.3   Contamination by Trace Metals

Concerns regarding possible contamination of aquaculture products by trace metals and organochlorines have arisen because of the poisoning episodes in many parts of the world, and the identification of several trace metals, dioxins, dibenzofurans and coplanar PCBs as the causative contaminants in aquatic environments. However, most of these have not been found in aquaculture products, and those that are found in any appreciable concentrations are not considered to be of public health importance.

The trace elements that are of greatest concern are mercury and cadmium. The Minimata disease caused by mercury contamination from discharges from a chemical factory, identified for the first time from Minimata Bay in Japan, has received considerable attention all over the world. Subsequently, a number of other mercury poisoning episodes due to eating contaminated fish have been reported.

As a result, several countries have introduced legislation concerning permissible limits of mercury in fish products. Internationally observed standards for trace elements in marine fish and shellfish are now available (Nauen, 1983). The median standard for mercury in fish is 0.5 μg g^{-1} wet weight for fish and shellfish (with a range of 0.1–1.0 μg g^{-1} wet weight). Such accumulations do not appear to have been reported from aquaculture products so far. Though mercury is usually accumulated in lethal concentrations by only long-lived predatory species such as tuna and swordfish, the possibility of smaller farmed fish accumulating it in significant quantities in areas of excessive local contamination cannot be ruled out.

Cadmium is another trace metal that is highly toxic to mammals, but is not known to have caused any public health problems in humans. However, there is significant concern about possible cadmium contamination due to the occurrence in Japan of the 'itai-itai' syndrome (an unusual disease of a rheumatic nature), believed to be caused by the discharge of cadmium-rich effluents from a zinc mine situated adjacent to a river about 50 km upstream from the affected area. The waters of the river were used for irrigation of rice fields, and consumption of rice from the affected area, containing high levels of cadmium, is thought to be the principal cause of the disease. The median standards laid down for cadmium in fish and shellfish are 0.3 and 1.0 μg g^{-1} wet weight, respectively. Though there are no recorded cases of cadmium poisoning in aquaculture products, there is a need to be vigilant, especially of the possible accumulation in soils of fish ponds fertilized with inorganic fertilizers containing cadmium. Oysters (*Crassostrea gigas* and *C. commercialis*), mussels (*Perna viridis*), clams and cockles, as well as certain species of crabs, can accumulate significant amounts of cadmium in polluted environments.

Another trace metal of concern as an environmental contaminant is lead, especially because of its possible effect on children. Concentrations of lead in seafood, however, are quite low and are unlikely to be a public health threat.

Copper and arsenic are also important contaminants of food products, but there appear to be no reports of any significant bioaccumulation of these in aquaculture products. Copper is of low toxicity to mammals and the arsenic found in most aquatic organisms is principally organic in nature. Concentrations of inorganic arsenic in aquatic species is very low (generally below 0.5 $\mu g\ g^{-1}$ wet weight) and are thus of little toxicological effect.

12.4   Contamination by Organochlorines

While organochlorines are known to be toxic to aquatic organisms and terrestrial animals, their direct toxicity to humans is low and there are very few known instances of their public health impacts. The few groups of organochlorines that are hazardous to humans include dioxins, especially TCDD (2,3,7,8-tetrachlorodibenzo-p-dioxins), dibenzofurans and certain polychlorinated biphenols (PCB). The well-known instances of contamination by dioxins, polychlorinated dibenzofurans and coplanar PCBs are the contamination of rice oil in Japan and Taiwan (Tanabe *et al.*, 1989; Chen & Hsu, 1986). The more widespread and disastrous environmental contamination that received international attention was the accidental emissions containing significant amounts of dioxin from a plant producing trichlorophenol in Seveso in northern Italy.

Although dioxins are an environmental contaminant, exposure most often occurs through food by consumption of animal fats. Unfortunately, there are no safe and effective treatments to remove the dioxins now in humans. Dioxins metabolize slowly (over years), but reduction in exposure reduces dioxin levels over time. The best way to reduce an individual's dioxin level (and risk from dioxins) is to reduce exposure and intake of dioxins. Recently, the European Commission has assessed the dietary intake of dioxins and related PCBs by the populations of EU member states. Some Baltic fish had concentrations that concerned local authorities and there is apprehension that fishmeal prepared from these species could bioaccumulate in animals fed with feeds incorporating these fishmeal products. From 1995 to 1999, the average concentrations of dioxins and related PCBs in UK salmon Baltic wild salmon and farmed salmon were 0.71, 7.04 and 1.47 (in pg TEQ g^{-1} product as toxic equivalents of 2,3,7,8-TCDD), respectively.

Anglers and fishermen in Nordic countries were noted to have higher concentrations of dioxins and PCBs than the general public, which was attributed to the consumption of the fatty fish. The Commission of the European Communities (2002) assesses acceptable levels of dioxins in feeding stuffs and foodstuffs with reference to background levels in order to prevent unacceptably high exposure levels among animals and the human population, as well as the distribution of feeding stuffs and foodstuffs with an unacceptably high contamination.

Shellfish such as mussels have been found to accumulate DDT and coplanar PCBs in significant quantities in highly polluted areas and can thus pose public health problems if such areas are used for shellfish farming. The effects of chronic low-level exposure to such contaminants in humans are not sufficiently well known. The US Environmental Protection Agency (US EPA 1999) has not issued a definitive statement concerning the hazards of TCCD, but in a draft EPA document, Guidelines for Carcinogen Risk Assessment, TCDD is suspected to be associated with key events in the carcinogenic process.

12.5 Microbial Contamination of Shellfish

In many countries, the growing of shellfish is not prohibited in areas subject to low-quality water with high bacterial concentrations. However, the quality of produce for sale may be regulated under sanitary laws. In some countries, as in Europe, the culture areas may be classified according to the bacteriological quality of the shellfish meats. Filter-feeding shellfish grown in polluted or low-quality water have to be depurated before sale, as they are known to accumulate pathogenic micro-organisms from contaminated waters.

Controlled purification is a well-established practice to protect public health, especially in areas where shellfish is eaten raw or partially cooked. Purification is necessary also to reduce the unknown hazards of viral pathogens, such as those causing infectious hepatitis, which are extremely difficult and costly to monitor in the shellfish or its environment. There are many instances of infectious hepatitis transmitted through oysters and clams.

The two main methods of purifying shellfish are relaying them in clean water or depuration in special plants (Alfsen, 1987). Relaying involves transplantation from an unsanitary area to a clean area for a minimum period, usually about 30 days. In France, oyster shippers are required to hold relaid oysters in special storage tanks filled with filtered and ultraviolet-treated seawater for another 2 to 3 days for self-purification before shipment.

Specialized depuration plants use chlorine or ozone for disinfecting seawater. The depuration process varies between plants, but the same basic principle is applied by all. Seawater is disinfected with a suitable disinfectant and used for filling tanks, where the shellfish remain for a specified period to allow the gradual release of pathogenic organisms through their characteristic pumping action. The water may be changed during the depuration process and an aeration system installed if required.

Packages of shellfish purified in approved plants are appropriately labelled for identification in markets. This has contributed substantially to the protection of public health in shellfish-consuming areas.

12.6 Contamination of Fish in Waste-water Ponds

The use of waste water and animal waste in fish farms has been described in Chapter 6, and concerns about the safety of fish raised in such farms have been discussed. Even though no public-health problems have been reported so far, apart from the controversial hypothesis of influenza pandemics from pig—duck—fish zoonesis in Chinese integrated farming (Schotissek & Naylor, 1988), the possibility of hazards continues to be raised, mainly because of successful experimental infections. Ponds fertilized with untreated sewage or other wastes can contain human pathogenic organisms. Laboratory experiments with tilapia, common carp and silver carp have shown that when bacteria are present in high concentrations in experimental ponds for long periods, the bacteria could be recovered from all organs and muscles of fish, although normally they are found only in the digestive tract and not in muscle tissues. Guidelines issued by the World Health Organization (WHO) for use of waste water (WHO, 1989) suggest that there is little accumulation of enteric organisms and pathogens on, or penetrating into, edible fish tissue when faecal coliform concentration in fish-pond waters is below 10^3 per 100 ml. When the concentration is greater than 10^4 or 10^5 per 100 ml, the potential for invasion of fish muscle by bacteria increases with the duration of exposure of the fish to the contaminated water.

The digestive tract and intraperitoneal fluid of fish may contain high concentrations of pathogens, such as salmonella, even at lower concentrations in pond water. Handling and cleaning of such contaminated fish can result in contamination of the hands of farm workers and through them, their family members and others.

Experimental work has shown that by appropriate treatment and management, including adequate retention time in oxidation ponds or lagoons, bacterial concentrations in waste water can be reduced and maintained at the required levels to prevent transmission of pathogens to consumers.

12.7   Contamination by Algal Toxins

The occurrence of red tides or algal blooms and their impact on aquaculture has been described in Chapter 7. These blooms can be toxic, rendering filter-feeding shellfish toxic almost overnight, or noxious, clogging the gills of filter-feeding fish and other animals, or cause oxygen depletion through decay and thereby anoxia. The toxic algae constitute major public-health hazards, besides causing mass mortality of fish and shellfish. Worldwide occurrence of toxic algal blooms has been reviewed by Shumway (1990).

Though historically the most common algal blooms were of dino-flagellates, more recently blooms of other species such as *Dinophysis*, *Aureococcus* and *Gymnodinium* have also become common. The toxins associated with these algae are highly potent, and shellfish feeding on them accumulate the toxins, becoming vectors of various forms of shell-fish poisoning, such as paralytic shellfish poisoning (PSP), diarrhoeal shellfish poisoning (DSP) and neurotoxic shellfish poisoning (NSP). Consumers of affected shellfish become poisoned and, therefore, marketing of shellfish is generally prohibited.

Highly successful shellfish farming in many areas has suffered heavily due to DSP and PSP. The toxins produced by dinoflagellates, which are water-soluble, consist of 12 sulfocarbamoyl and carbonate toxins. Enzymatic action on these toxins may give rise to decarbamoyl toxin derivatives (Sullivan *et al.*, 1983; Shimizu & Yoshioka, 1981), sometimes increasing the number of toxins to 18.

The dinoflagellate *Protogonyaulax tamarensis* is the species that has caused PSP in many parts of the world, although other species have also been reported (see Chapter 7) from Europe and the northwest Pacific. As crustaceans do not accumulate these toxins, they are safe to eat even during intense blooms.

NSP, which has milder symptoms similar to PSP and is caused by *Ptychodiscus* (= *Gymnodinium*) *breve*, has been reported to occur annually along the coasts of Florida and Texas in the USA. DSP has symptoms very similar to gastroenteritis associated with eating shellfish, and can, therefore, go unidentified. Toxins associated with DSP include okadaic acid dinophysistoxin-1 and dinophysistoxin-3 (Lee *et al.*, 1989), which are lipid-soluble. Recent research has demonstrated multiple toxin profiles associated with DSP (Lee *et al.*, 1988).

While *Dinophysis acuminata* is considered to be the main source of DSP in Europe, *D. norvegica* and *D. acuta* are reported to be the most likely organisms that cause DSP in Norway. *D. fortii* is also believed to be implicated in DSP occurrence in Japan, and the benthic dinoflagellate *Prorocentrum lima* in Spain.

It is important to note that a bloom is not always necessary to induce toxicity, and even low concentrations of 100 cells l^{-1} may produce high levels of DSP in mussels (Dahl & Yndestad, 1985). The sudden and unexpected occurrence of amnesic shellfish poisoning (ASP) in Prince Edward Island (Canada), caused by the toxin demoic acid believed to be derived from the diatom *Nitzchia pungens*, has been referred to in Chapter 7. Shumway (1990) lists a number of additional algal species that have been implicated in shellfish poisoning.

Algal toxins persist for many months after the blooms have disappeared, and can, therefore, cause major economic damage to shellfish culture industries, besides causing public-health problems. As the algal blooms originate in the oceans far from aquaculture sites, and are caused by climatic and other factors that are not fully understood, there is very little that a farmer can do to prevent their occurrence. The farmer's main aim has to be to protect the stock and to avoid marketing contaminated products. Closure of farms during periods of the year when algal blooms occur, and the establishment of shellfish toxicity monitoring programmes, are considered to be the most effective means of achieving this aim.

Uptake of toxins by shellfish varies with species, intensity of blooms and water temperature. Mussels are known to accumulate PSP toxins faster than others, and also to eliminate the toxins quickly. On the other

Figure 12.1 Depuration of shellfish in France. (*Photo by H. Grizel*)

hand, oysters take a longer time to accumulate toxins and also to detoxify. Some species are reported to remain toxic for long periods, even in excess of two years (Quayle, 1965; Blogoslawski & Stewart, 1978).

Luckily for the farmer, there are some species that can avoid toxic algae. The hard clam or quahog *Mercenaria mercenaria* has been observed to close its shell and isolate itself or bury deep when exposed to toxic algal blooms, and thus avoid becoming toxic. Scallops are similarly safe to be marketed during periods of toxic algal blooms as the abductor muscle, which is the part usually eaten, rarely, if ever, becomes toxic (Shumway *et al.*, 1988).

Depuration of contaminated shellfish by holding them in waters free from toxic organisms (see Fig. 12.1) is feasible for many species, but some species remain toxic for extended periods. Other methods of detoxification using temperature or salinity stress and chlorination have been tried, with varying degrees of success. Chlorination, though effective, changes the flavour of the shellfish, making it unacceptable in many markets.

A more successful means of detoxification appears to be ozonation, though there are different views on its usefulness and observed limitations to its effectiveness. Based on a review of results so far, Blogoslawski (1988) concluded that ozonized seawater can be valuable in detoxifying shellfish recently contaminated by the vegetative cell phase of toxic dinoflagellates. Ozone treatment of seawater has been shown to prevent accumulation of paralytic shellfish toxins in *Mytilus edulis*. However, ozone cannot detoxify shellfish that have ingested cysts or have toxins bound in their tissue for a long time.

Taking into account the findings so far and the economic considerations, it has to be concluded that detoxifying shellfish on a large scale is not particularly promising at present (Shumway, 1990). Under these circumstances, the best solution appears to be an intensive and reliable monitoring programme in areas prone to regular outbreaks of algal blooms, and restricting culture to species that detoxify rapidly, such as the mussels, and species known to be free of toxins, such as the hard clam (quahog) or scallops, where the edible parts are free of toxins.

Though prediction and early warning of toxic algal blooms would be the ideal means of avoiding major economic loss to shellfish growers, and preventing poisoning of shellfish consumers, there is as yet not enough information on the causes of such blooms to enable reliable predictions. In the absence of an early warning system, monitoring is considered to be the only available management tool. Many countries have established comprehensive monitoring programmes, consisting of monitoring hydrographic conditions, phytoplankton production (including detection of accidentally introduced hazardous species) and toxin monitoring in shellfish. Many of these have paid off by enabling early action to prevent disastrous losses to growers and the occurrence of shellfish poisoning among consumers.

Chapter 13
Sustainability of Aquaculture

The concept of sustainability is one of the factors that have influenced the progress of certain forms of aquaculture in recent years. The Earth Summit in Rio de Janiero in 1992 resolved that all forms of development, particularly those utilizing natural resources, should be based on the principle of sustainability. This happened at a time when many of the World's fishery resources were showing signs of over-fishing, yielding diminishing returns. As mentioned earlier, aquaculture became the only growth sector of world fisheries. The possibility of establishing aquaculture-based fisheries became an acceptable activity to be considered. Development planners concluded that within the next two decades aquaculture and culture-based fisheries may be able to meet more than half the World demand for fish and shellfish. This has been supported by the yearly percentage increase in aquaculture production in most countries. However, to achieve the potential production, it has become necessary to follow sustainable development policies and changes in aquaculture technologies.

In the past, research and experimentation have been guided by the objective of obtaining increased yields by intensifying aquaculture practices. Lack of tested sustainable practices was viewed as another impediment to the emerging infant industry, without a clear idea of the dimensions of sustainability. Aquaculture has been based on the principle of short-term economic viability. When this was affected by disease outbreaks as a result of self-pollution or external waste discharges, there was general recognition that environmental sustainability is a valid idea which must be considered.

13.1 Definition of Sustainability

Several national and international bodies have been trying to define sustainability with respect to developments in various fields, especially those that make use of natural resources, like agriculture, animal husbandry and forestry (Pillay, 1997). From among the definitions that describe the impacts on the natural resources of land and water, the most commonly accepted one, which is applicable to aquaculture, is the one that was developed mainly for agriculture, forestry and fisheries at the FAO (1991) Den Bosch, The Netherlands, conference. This defines sustainable

112

development as the management and conservation of the natural resource base and the orientation of technological and institutional change in such a manner as to ensure the attainment and continued satisfaction of human needs for the present and future generations. Such developments conserve land, water, plant and animal genetic resources, in environmentally non-degrading, technically appropriate, economically viable and socially acceptable manner. This definition concerns not only the present but also future generations, and the conservation of natural resources and genetic diversity. It also refers to the appropriateness of technology and economic viability as well as social acceptability, which are of prime concern in sustainable development.

Among the other definitions, the one that needs attention from the point of view of aquaculture is that proposed by the World Resource Institute, which describes sustainability as a system that improves in a substantial and enduring way the underlying productivity of natural resources and cropping patterns, so that farmers can meet increasing levels of demand in concert with population and economic growth, as well as environmental necessities.

The Bruntland Commission (the World Commission on Environment and Development), in its report on *Our Common Future*, defines sustainable development as that which meets the needs of the present, without compromising the capacity of future generations to meet their own needs. The two major concepts contained in this definition are the concept of the essential needs of the World's poor, and the perceived limitations imposed by the state of technology and environment to meet future needs.

From the definitions quoted it would appear that the prerequisites at present are the controlled use of natural resources on a renewable basis to meet the food security of increasing populations and their economic growth. Preservation of options available to future generations is highly predictive in nature and can be attempted only with great care on the basis of trends in socio-economic growth. There is every likelihood that advancement of technology will accompany generational changes and it is difficult to foresee the interdependent options available to future generations. One can foresee the environmental management, but at the same time, it is not likely that a generation can avoid using intensive production technology to feed surging populations and, in turn, cause greater environmental perturbation. With the efforts to eradicate poverty, succeeding generations may not be so poor as now, and their needs may change in keeping with future economic development.

13.2 Economic Sustainability

Farmers, producers and investors have found aquaculture an economically profitable vocation when compared with fishing, which provides only diminishing returns. Export possibilities, added to the short-term profits of aquaculture of species like shrimps, trout and salmon, have the potential for earning foreign currency. Returns from investments underwent considerable changes as a result of market fluctuations and overproduction. Intensive farming gave rise to diseases, which reduced yields and consequently returns on investments. On the whole, growth in farming was not directed to sustainability on a long-term basis and did not take into account natural-resource assessment that incorporates environmental externalities in cost–benefit analyses. Nor did the industry try to involve local communities or to create a public image that could have helped to resist the opposition of environmentalists, media and politicians.

The involvement of companies and large corporations gave rise to questions about the equitable distribution of benefits from farming. It was pointed out that in shrimp farming in Ecuador, which produces more than 60% of its shrimp exports from aquaculture ponds, benefits are confined to a limited number of entrepreneurs, government officials and foreign experts (Landesman, 1994). This is believed to be the case in many Asian countries as well. Some investors tried to share the benefits among local populations through contract farming, or social-benefit activities such as providing drinking water supplies, or assisting in the provision of seedlings from hatcheries and marketing of farm products, or the distribution of inputs. However, this did not prove sufficient to prevent resentment of the general public, probably due to lack of mutual trust.

13.3   Environmental Sustainability

Environmental impacts of aquaculture are very much associated with the type of farming adopted and the species under cultivation. The sites where farms are located have a considerable role in determining the environmental impacts of culture operations. This is why site selection for aquatic farms is considered to be of significant importance in the sustainable management of aquaculture. Site selection has to be based on the assimilative capacity (GESAMP, 1986; GESAMP, 1991) for the estimated amount of effluent that is discharged into the neighbouring waterways and the ability to disperse it far away from the farm. Estimates of waste production have been reviewed in Chapter 7. As can be seen from the data presented, most of the investigations reviewed relate to temperate regions and to salmonid culture. Since many of the present-day problems have risen from shrimp farming in the tropics, it may be of interest to quote here the relevant data (Table 13.1) from Thailand detailed by Barg (1992).

The quantity of wastes from all types of aquatic farm will vary with the intensity of farming operations, but the assimilative capacity of waste discharges will depend very much on the flushing rate of the receiving waters or the regular removal of farm sediments for their disposal. Since many farms are situated near water inlets, it is considered

Table 13.1   Effluent water quality from an intensive shrimp farm in Thailand during a five-month grow-out (Barg, 1992).

Variable	Value	Units
Pond size	0.48–0.56	ha
Pond depth	1.50–1.80	m
Salinity	10–35	psu
Temperature	22–31	°C
pH	7.50–8.90	
Total phosphorous	0.05–0.40	mg l^{-1}
Total ammonia	0.05–0.65	mg l^{-1}
Dissolved oxygen	4.0–7.5	mg l^{-1}
Chlorophyll *a*	20–250	mg l^{-1}
Total suspended solids	30–190	mg l^{-1}
Water exchange	5–40	% day^{-1}

beneficial to have sedimentation tanks associated with inlets of farms, though their usefulness has not been scientifically proved in each case. Where regulations have been established, one important condition to be satisfied is to reserve enough space for settling tanks, usually about 10% of the farm area. This may affect the economics of farming, especially in areas where suitable sites are scarce and costly. Negroni (2000) considers constructed wetlands are an attractive option for the disposal of fish farm effluents. Macrophytes can clean waste water of potential pollutants by direct assimilation. The major removal mechanism for nitrogen is nitrification and denitrification. Phosphorous removal occurs as a result of adsorption. Pathogens are removed during passage of waste water through sedimentation and filtration facilities. However, based on experiments, Schwartz and Boyd (1995) concluded that this is not feasible in commercial catfish farms.

13.4   Social Aspects of Sustainability

Sustainability of aquaculture development is very much dependent on effective management of social impacts. Insufficient involvement of local communities often gives rise to crisis situations and builds up avoidable resentment. There is a need for ensuring equitable distribution of benefits to local communities, especially by larger enterprises owned by non-indigenous companies and corporations, as this can often be the cause of opposition to aquaculture development in many areas. Most small-scale aquaculture in developing countries is undertaken by farmers or small fishing families as an alternative means of livelihood, where the major consideration is whether the opportunity is fully recognized and the time and effort expended are considered worthwhile for maintaining a standard of living acceptable to the community. In contrast, large-scale aquaculture enterprises are established as investment diversification, to obtain maximum returns on investments. If for some reason these do not succeed, the owners more often than not abandon the projects, as has happened with abandoned shrimp farms in reclaimed mangroves.

Social problems can arise if large-scale farms are established on agricultural land, even if they are low-yielding, and bought from owners at much above the prevailing prices. When considering the impacts of aquaculture on the present and future generations, predictions have to be based on trends observed. Controlled use of natural resources on a renewable basis to meet the needs of present and future generations may not be very easy. Employment opportunities opened up by the application of new scientific technologies and educational opportunities for future generations have to be considered. Aquatic farmers and fishermen in developing countries are among the poorest of the poor, but technological advances can bring about unbelievable changes in their habits and aspirations, as already shown by their adoption of information technology in the day-to-day management of individual farms, and marketing of aquaculture products. This shows that predictions can be made only on a short-term basis.

It has been pointed out earlier that site selection for farm construction has a significant social impact. It is believed that aquatic farms, if not properly located, can affect the present livelihood of neighbouring villages. Very often coastal aquaculture farms prevent easy access to beaches where small-scale fishermen keep their boats and dry their nets. Farms may obstruct small-scale fishermen from carrying out their fishing activities. Complaints about attempting to privatise common property resources have to be avoided. Similarly, if proper care is not taken in the construction of farm embankments, it is likely that neighbouring agriculture fields in coastal areas may be affected by salinisation of soils. Drinking water sources may also be salinised by farms located in coastal areas. Where ground water has to be pumped for reduction of salinity in farm ponds, competition may arise in the use of fresh water, and there is also the risk of subsidence of land. As noted earlier, farm construction in coastal areas may give rise to soil erosion and destruction of mangroves.

Large farms that grow exportable species often undertake monoculture that does not involve integration with crop and animal husbandry, thus preventing diversification of food production which would spread the farmers' risks. Aquaculture operations may reduce opportunities for off-season work for local populations if they cannot be involved in seed collection for sale to producers.

Much of the opposition to aquaculture may be due to lack of information on the possibilities of undertaking sustainable aquatic farming. There is, therefore, a need for disseminating authentic information based on actual experience, and the results of research and experimentation on sustainable aquaculture.

13.5   Guidelines for Sustainable Aquaculture

Even though there is general agreement on the need for aquaculture and culture-based fisheries to meet the increasing demand for aquatic products, there is considerable controversy about the procedures to be followed for achieving the inter-generic and social dimensions of sustainable aquaculture development (Pillay, 1996). The absence of adequate scientific information cannot be used as a reason for postponing or failing to take conservation and management measures (FAO, 1995). Farmers and producers have acquired experience in modified semi-intensive polycultural practices, which can be used for focusing future research, and guide present-day farmers and producers to follow procedures that are likely to become technologies to achieve sustainability.

Recognizing the need for codifying and analyzing the experience gained so far, and the results of studies and experimentation, several national, regional and international institutions took initiatives to collate existing data and prepare guidelines or codes of conduct for the future (FAO, 1995; INFOFISH, 1997; NATS, 1998; FAO/NACA, 1995; D'Abramo & Hargreaves, 1997; ADB/NACA, 1998). The Choluteca Forum (1996) of non-governmental organizations that considered shrimp farming in tropical and subtropical areas took a different approach and demanded the imposition of a global moratorium to halt further establishment or expansion of shrimp farming in coastal areas until the criteria for sustainable aquaculture are put into practice. The guidelines relating to sustainable aquaculture based on existing information are summarized below. This should be read along with Section 9.4 on control of introductions and Section 9.6 on guidelines for management of movements of live aquatic animals.

The guidelines prepared through various consultations and conferences and international organizations set out principles and standards to be followed for ensuring effective ecological sustainability. Even though it is recognized that sustainability can be achieved only with the co-operative efforts of farmers and producers, the governments have a major role in promoting and facilitating sustainability through enacting legislation and support for its implementation. Therefore, most of the guidelines are addressed to sovereign-state governments. The most important principle concerned in the guidelines is planning of aquaculture development on the basis of availability of resources and zoning of suitable areas without conflict with traditional uses as far as is possible. This may involve balancing benefits and detriments, including quantification of economic, environmental and social aspects. All planning of farming should be preceded by impact assessments and formulation of measures mitigating against adverse impacts. Regular monitoring of these measures has to be promoted to minimize adverse ecological changes, and the rational use of resources shared by aquaculture and other legitimate activities (Roberts & Muir, 1995; Hopkins, 1996).

Even though the guidelines or codes of conduct are not mandatory, regular legislative measures, such as licensing of farms based on impact assessment, can assist the implementation of the guidelines which are meant to facilitate long-term economic benefits, and lead to sustainable aquaculture practices for present and future generations. These can enforce participatory planning of development in support of rural communities, producers and fish-farmers. When designing and constructing farms, the livelihood of local communities and their access to fishing grounds must not be impaired. Use of groundwater resources to reduce salinity, which would result in subsidence of earth, should be avoided, as should salinization due to seepage from the farm, which might affect adjacent agricultural land. Development agencies should promote effective farm and fish health management practices, including use of vaccines and other hygienic measures. Safe and effective use of therapeutics, hormones, drugs, antibiotics and other disease control chemicals should be ensured. It is necessary to regulate the use of chemicals that are hazardous to human health and the environment. State agencies concerned should ensure the food safety of aquaculture products, including genetically modified food organisms, and help to improve and maintain their quality.

Even when farms are designed and operated according to scientific principles, clustering of sites without due account of their carrying capacity can damage the environment and affect their sustainability. Guidelines relating to coastal aquaculture and culture-based fisheries within trans-boundary aquatic ecosystems emphasize the need for integrated coastal zone management. It is important that areas and resources important for various types of aquaculture are protected from being irreversibly allocated for other purposes. Co-ordination between relevant government departments has to be enforced to facilitate the implementation of guidelines. As emphasized in Section 10.6, regulations regarding movement of exotic and genetically modified species can be implemented only with the willing cooperation of the farmers, producers and local government officials. It should be the responsibility of the appropriate authority of the state agencies to protect trans-boundary aquatic ecosystems by promoting sustainable aquaculture practices within their national jurisdiction, with the cooperation of the neighbouring states, in accordance with international law. The choice of species and siting of aquaculture activities that could affect trans-boundary aquatic systems should be decided on the basis of consultations among state agencies concerned. Suitable mechanisms such as data bases and information networks have to be established to collect, share and disseminate data relating to aquaculture and aquaculture-based fisheries, in order to facilitate co-operation in planning and development at national, regional and global levels. Every effort should be made to conserve genetic diversity and maintain integrity of aquatic ecosystems. Steps should be taken to minimize disease and other detrimental effects caused by fish which escape from farms and from enhanced stocks.

Where states do impose licensing of farming enterprises based on environmental impact assessments, the licenses should include compulsory monitoring of the results of mitigation measures, including water usage and waste-disposal requirements such as provision of settling tanks, restrictions on use of mangrove swamps, and polyculture with bivalves and seaweeds that make productive use of suspended and dissolved wastes. Since much of the waste produced in fish culture farms comes from feeds and feed spillage, feed manufacturers have to be encouraged to improve feed quality by reducing the protein and increasing the lipid content to decrease the emission of nitrogen into the environment. Processing of feeds by the extrusion method ensures production of floating pellets. Even though these feeds are more expensive, such eco-friendly feeds should be recommended wherever possible and particularly in large-scale farms. Because of the predicted scarcity of fish meal for fish-feed manufacture, the scientific community should give priority to the development of sources other than fish protein and fish lipids as ingredients in the formulation of feeds for aquaculture. In recognition of the potential of aquaculture to contribute significantly to the World's aquatic food supply, national, regional and international organizations should give high priority to the transfer, adaptation and implementation of technological innovations and capacity building, in order to stimulate sustainable aquaculture practices and establishing economical culture-based fisheries.

Chapter 14
Economics and Environmental Impact Assessments

14.1   Development Planning and Public Information

National aquaculture plans, where they exist, are mainly directed to establishing or strengthening aquaculture through cost-effective production and marketing techniques, and meeting domestic and export requirements. The environmental impacts of the industry, which were often given short shrift, became a matter of major concern only with the intensification of culture technologies and increasing utilization of coastal areas for pen and cage farming. Solutions to problems were sought only when emergencies such as serious local conflicts or disastrous mortality or toxic poisoning occurred. Experience has shown the need for prior planning to ensure the sustainability of aquaculture enterprises and avoid environmental degradation.

The previous chapters have highlighted the scarcity of adequate information on a number of environmental aspects of aquaculture systems. Acquiring such knowledge, which requires considerable research, investigation and development of environmental expertise, will obviously take time. Development of aquaculture in a country cannot be halted until all the necessary information has been acquired, and so planning has often to be based on available, albeit incomplete, data or appropriate analogous information obtained elsewhere. Planning is a dynamic process and all plans require revision or modification when additional information and experience are gained during their implementation. Environmental assessment and monitoring should, therefore, form an integral part of an organized development plan.

Planning is required at a national macro-level, as well as on the project or farm-based micro-level. Rationalization of access to land and water resources, selection of sites and farm design, reconciliation of conflicts and harmonized resource use, appropriate farming systems, product quality, and legislative support for effective monitoring and management are central to such planning. As environmental concern, or its absence, and conflicts with other users of the resources are often based on lack of information or on misinformation, a well-organized public information system should form an integral part of a national plan. In some areas there is a need to bring about awareness among the public and others concerned that aquaculture is a legitimate activity, serving national and social goals.

The profit motive of a farmer or an entrepreneur need not be considered as inimical to the concepts of social justice and sustainability. Planning at the farm or enterprise level would normally require some specialized expertise and knowledge of the environmental consequences of the proposed farming system in similar site conditions. Appropriate guidelines issued by agencies authorized to approve aquaculture projects would enable project sponsors to determine in advance the procedures involved in obtaining necessary permissions, especially whether the proposed project requires environmental clearance. If such clearance is required because of the size or nature of the project, the guidelines should specify whether the initial application for permission should be accompanied by a preliminary environmental assessment following a prescribed format, in order to determine the need for a subsequent detailed assessment for project approval.

These procedures are described in more detail in later sections of this chapter. The concept of integrating environmental impact assessment and reduction or avoidance of adverse effects into the overall project activities, from design to implementation and monitoring, has the potential to allay suspicions of incompatibility of environmental concerns with productive aquaculture.

The nature of information that a regulating authority would normally require is related to the site selected, the proposed farming system, the ecological and socio-economic conditions of the area and possible conflicts with other enterprises in the neighbourhood. These have been discussed in some detail in Chapter 5. As discussed in Section 13.5 on sustainable aquaculture, the project plans should include predicted consequences of the farming and related activities on the environment, and methods proposed for mitigating adverse impacts as well as enhancing the beneficial impacts.

The descriptions of the nature and extent of impacts in Chapters 3 and 4, and the discussions on their causes and mitigating measures in Chapter 5, provide the available basic data for such predictions and management procedures. However, the applicability of the available information to the project being planned has to be determined by on-site investigations.

While by proper planning and management many of the environmental impacts can be avoided, there are inevitable changes that will occur, and unfavourable changes have to be kept at acceptable levels. As is evident from earlier discussions, the scientific base for development of mitigation measures is extremely weak at present and greater research effort is a priority, both nationally and internationally. Research on aquaculture technologies and systems should include, as a matter of necessity, the environmental consequences of their adoption, including the socio-cultural and economic effects.

14.2   Aquaculture Development Zones

Selection and acquisition of sites probably form the most important basis for conflicts in aquaculture development, and the nature and operation of the farming system constitute the major cause of environmental concerns.

With the increasing importance of the aquaculture sector in most countries of the world, there is a valid case for macro-surveys of the ecology, including social and economic conditions of areas that are suitable for the introduction of selected culture systems that could meet national or the introduction priorities of development. Such surveys and studies could form the basis for decisions on allocation of publicly owned land and water resources for aquaculture and other uses. The allocations would have to be made jointly by the governmental agencies concerned under the technical leadership of the agency responsible for aquaculture development, and in consultation with local authorities and non-governmental organizations.

Provision has to be made for the conservation of areas of coastal marshes and mangrove swamps to enrich estuarine and marine environments adequately and preserve optimum populations of birds and other wildlife.

It is recognized that all the basic information required for the necessary estimations is not available, but it is possible to make approximate calculations based on existing reports and extrapolations. As a precautionary measure, allocations of new activities can be made at a conservative level. As will be clear from earlier discussions, such allocations and general guidelines could be of considerable help in streamlining procedures for granting permission to own or lease suitable sites. Public hearings may still be needed on a local level for granting permissions, but appropriate guidelines and adequate public information should reduce procedural delays in reaching decisions.

Such general allocation of land and water for aquaculture should take into account the farming systems that can be practised in the area, the maximum area that can be occupied by an enterprise, the maximum production admissible in each production unit in relation to waste disposal requirements, and the sanitary measures to be adopted. Applications should be supported with all available data relating to the site, the culture system to be employed, the likely social and environmental impacts and measures proposed to mitigate unfavourable effects. The use of environmental impact assessment (EIA) for this purpose will be required in order to protect the environment and the operation of other enterprises in the neighbourhood. It will be necessary to enforce

Figure 14.1   Salmon cage farms located in 'weir' fishing grounds in the Bay of Fundy, Canada. (Photo by D.E. Aiken)

these requirements in respect of both privately owned and state-owned sites.

Conflicts with traditional fishing activities in foreshore areas and estuaries may figure prominently in the selection and acquisition of sites for aquaculture in certain areas, especially when farming is to be carried out by non-local enterprises. Management problems created by salmonid aquaculture in the Bay of Fundy, Canada (Fig. 14.1), in areas of traditional fishing for herring, lobster, clams and scallops, is an example of such conflicts (Stephenson, 1990). Considering the nature and extent of impacts, the suggested solution is the development of a zoning system which would provide adequate protection to important fishery areas, while allowing multiple use, including aquaculture.

Another source of conflict with capture fisheries is the collection of larvae or juveniles of aquaculture species for rearing, or of brood fish for spawning in hatcheries. While no adverse effects on wild populations have been observed, as mentioned in Chapter 1, the best solution is to depend on hatchery production of seed and controlled maturation of brood fish.

If Barbier's (1987) criteria for sustainable development are followed, intensive aquaculture systems have to be rated as less favourable than extensive ones. Selection of farming systems, however, is governed by a number of considerations, but from the point of view of environmental impact, pond farms following semi-intensive or extensive systems have

Figure 14.2   A pond farm in China that merges well with the landscape.

many favourable features, although they tend to occupy more land area. When properly designed and maintained, they seem to merge well with the landscape and in many cases add to the scenic beauty of the countryside (Fig. 14.2). Temporary farm buildings can be designed to harmonize with the landscape, and outdoor storage and other installations can be adequately screened.

With generally longer retention time of water in ponds, the impact of waste discharges tends to be minimal. Demand for water can be reduced by preventing seepage by reinforcing dikes and pond bottoms. Sedimented organic matter within the ponds can be removed and used as crop fertilizer after the ponds are dewatered, or allowed to mineralize by simple means. Polyculture of species combinations with compatible feeding habits can help in maximizing primary productivity and reducing nutrient discharges. It may be possible in suitable situations to use minimum water exchange systems successfully (Avnilmech, 1998) in intensive fish farming.

Many of these advantages, from an environmental view point, are lost in intensive pond, tank and race-way culture systems, which maintain dense stocks with intensive feeding and less water retention time, causing comparatively larger waste production. However, in land-based systems and hatcheries, it is possible to employ suitable waste treatment methods, the additional costs involved being compensated by the reduced space requirements and the higher rates of production obtained.

The culture systems that have caused most environmental concerns are cage and pen culture in open waters and shellfish culture in coastal areas. Because of this, there have been relatively more scientific studies on the environmental impacts of these systems. The quantitative data now available on salmonid species are of considerable value in directing some remedial measures and future research. Wider dissemination of existing information should go a long way respond to environmental concerns, as the data indicate that the impacts are minimal when the farms are properly sited and operated.

The most important factor to be considered is the flushing rate of the receiving water and its capacity to handle the wastes and accumulated sediments from the farms. Though a cage culturist generally looks for protected bays and fjords for siting the farm, slow-flushing and semi-enclosed waters are prone to waste and sediment accumulation. Areas with stagnant water should, therefore, be avoided. One of the main reasons for the current efforts to develop offshore open-water cage farming is to reduce the impact of wastes, which not only affect the environment but also endanger the culture operations.

In the absence of reliable carrying-capacity models, over-crowding of sites has to be avoided by limiting farm size, stock magnitude and lay-out and moorings of cages, on the basis of available scientific information on waste production, sedimentation, nature of benthic communities in the area and seasonal flushing rates. The basic data for designing necessary procedures can be obtained through well-planned environmental assessment studies before farm construction. Regular monitoring will be needed after farming has started to determine the appropriateness and efficiency of the measures adopted, and the need to modify them.

14.3   Environmental Impact Assessment

The use of environmental impact assessment, which involves the analyses of potential interactions of development projects with environmental quality, has been in practice since the 1970s in countries such as the USA for all major development activities and selected environmentally sensitive projects. The pioneering legislation in this field was the National Environment Policy Act (NEPA) of 1969, requiring a systematic inter-disciplinary evaluation of the potential environmental effects of all major federally funded projects.

Subsequent growth of the EIA concept and related legislation in other parts of the world have not always followed this pioneering example, but were designed to fit into the constitutional, economic, social and technological framework of each country. Even terms such as environmental impact assessment (EIA) and environmental impact statement (EIS) may have different connotations. For example, EIA could also be used to denote a brief examination conducted to determine whether or not a project requires an EIA based on a set of guidelines that identify projects for which a full environmental study would be required. If the project is not exempted by the guidelines, a full environmental study is undertaken and the findings of the study are reported in an EIS.

Currently, in most countries outside the USA, the term EIA is used to 'include the technical aspects of the environmental study, including data gathering, prediction of impacts, comparison of alternatives and the framing of recommendations' (Ahmad & Sammy, 1985). If the term EIS is used, it refers to the document that summarizes the results of the study, and the recommendations forwarded to the decision-maker.

In the most widely accepted sense, EIA is a decision-making tool that constitutes a study of the consequences of a proposed action on the environment, comparing various alternatives and identifying the one that is the best combination of economic and environmental costs and benefits. It is based on the prediction of changes in the environmental quality that would result from the proposed action, and an assessment of environmental effects on a common basis with economic costs and benefits. It also examines how the project would benefit or harm the people, their homeland or livelihoods, or other nearby developments.

Identifying measures to minimise adverse effects of the proposed activity and ways of improving the project's suitability for the environment form the logical follow-up of the prediction of potential problems.

Developers do not always look very kindly on the need to undertake an EIA for a number of reasons, including the perception that it is poised against development, that it is expensive and results in very considerable delays in starting development projects, and that it is not adequately used in decision-making. While some of these perceptions may be applicable in the case of aquaculture enterprises, the major reason seems to be the belief, if not the confidence, that properly organized aquaculture causes very few environmental problems, and the impacts observed so far are only of a short-lived and local nature.

The scarcity of scientifically verified information on different potential adverse impacts of aquaculture have already been pointed out, and the fact that many of the concerns expressed are of a speculative nature does not necessarily mean that adverse effects will not be discovered when appropriate studies are carried out. In any case, many environmental regulatory agencies and financing institutions do now insist on EIAs for large-scale aquaculture projects, mainly to ensure that appropriate mitigatory measures are incorporated in the project design if adverse effects are expected.

The information requirements proposed by the Environmental Protection Agency of the USA for the approval of permits for discharges from aquaculture projects under the *Federal Pollution Control Act* (Sec. 318.501) have been referred to in Section 5.1. Additionally, it will be in the long-term interests of aquaculture enterprises for the assessments to be made, even if they are not mandatory.

Flexible methods of assessment, which can be carried out with available information and limited expertise and expense, have been recommended by the United Nations Environment Programme (UNEP) (Ahmad & Sammy, 1985; UNEP, 1988 and 1990). Essentially, such methods depend on the use of observed environmental effects in analogous cases, followed by monitoring of environmental impacts during project implementation as an integral part of the EIA process. This allows for errors in the initial predictions to be observed and corrected as more knowledge about environmental consequences becomes available.

As will be discussed later in this chapter, there is a need for appropriate institutional arrangements, backed by suitable legislation, to enable EIAs to be undertaken as a part of major development activities. It is only natural that such a requirement is linked to the granting of permits for the implementation of projects or activities, notwithstanding the usefulness of such assessments even in activities that do not require formal permits.

The responsibility for impact assessment is generally that of the person, private company, or government agency that undertakes the activity. However, the authority that is responsible for issuing permits or project clearances has an important role in the conduct of the assessments by providing appropriate guidelines and ensuring that agreed mitigation measures are implemented if the project is approved and becomes operational.

If it is a major project, there may be various groups that will be interested in its impact, and it is essential, therefore, that their concerns and points of view are ascertained and fully taken into account in the assessments. The results of the studies should be accessible to the groups concerned as well as to the competent government authority which issues the necessary permits and the financing agency. The local community needs to know how the project will affect them and their quality of life, and the politicians are likely to be interested in knowing who is affected and in what way, and what issues are of political significance. Concerned government authorities would want to know how other potential projects in the area will be affected and whether the impacts will interfere with adjacent developments and land uses.

Environmental impact assessment is a growing technology and has undergone changes in its concepts and operation. From being a legal requirement for obtaining approval for a developmental project, EIA is now considered an integral part of a project, from the origins of the project idea and pre-feasibility studies to its successful implementation, monitoring and evaluation.

Figure 14.3 illustrates the integration of EIA with other project activities in order to provide timely environmental information for improved planning and execution. This underlines an important principle of EIA, namely that of conducting environmental assessments to support directly the various stages of project planning and implementation.

It is important to restrict assessments to the most serious and likely impacts, so as to enhance the value of the exercise for decision-making and to economize on time, effort and money spent. The involvement of

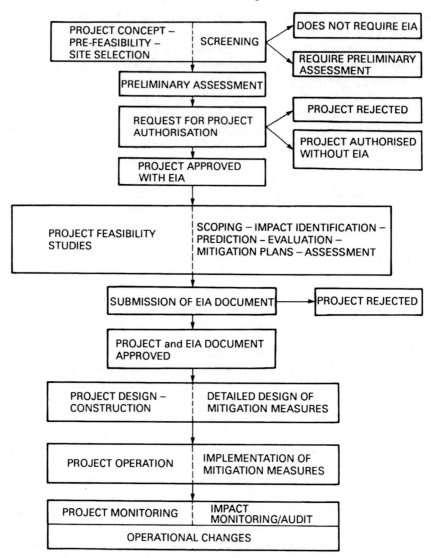

Figure 14.3   Simplified flow diagram, showing integration of project activities with environmental impact assessment (EIA).

persons or groups who can contribute, or are directly concerned or affected by the project, is essential to the success of EIA.

14.3.1   *Assessment procedures*

Not all development activities require detailed environmental assessment, and only major projects that are likely to have serious impacts should need detailed assessment. Screening and a preliminary assessment will be required as a first step at the project's pre-feasibility study stage to help in determining the follow-up actions.

Screening is generally done at the concept stage of a project to evaluate it against simple criteria, such as its size and location, and to compare it with the observed impacts of similar projects and set thresholds. This helps to clear projects that have been found, from past experience, to cause no serious environmental problems, for example, in some countries small-scale aquaculture producing 50–100 t of fish annually in pond or cage farms does not require a full EIA.

If the screening does not result in clearing the project, a preliminary assessment may have to be carried out by the developer. This would be directed to describing the nature and extent of the predicted key impacts of the development on the local environment, and evaluating their importance. If this evaluation shows that the project is unlikely to have any significant environmental impact, or that possible adverse effects can be minimized by immediate incorporation of mitigatory measures or modification of project design, it may be cleared by the competent authority without further assessment.

However, if the project is approved subject to a detailed EIA, the developer will be required to organize the required study according to prescribed guidelines, with the help of a group of specialists. The main steps involved can be described as: (1) scoping and impact identification, (2) base-line studies, (3) prediction and evaluation, (4) identification of mitigation measures, (5) assessment or comparison of alternatives, and (6) impact monitoring and auditing. These are carried out in concert with the technical and economic feasibility studies of the project to enable viable alternatives to be recognized.

14.3.2   *Scoping and impact identification*

The scoping of an EIA has the main purpose of ensuring that the study covers all the issues of importance in making decisions on the environmental sustainability of a development project, and also that it avoids gathering unnecessary data. It involves the listing of all the potential impacts of the proposed development by synthesis of known impacts from analogous cases. It should include not only the immediate and direct consequences on the project site but also the indirect consequences that can occur later or in other parts of the environment 'downstream'. Scoping should include both the negative aspects of development and opportunities for environmental enhancement offered by the activity. Socio-cultural and economic consequences are especially important in aquaculture projects.

This checklist is examined for the magnitude of change that can be expected, the extent of the area that will be affected, the significance of the change and whether the expected changes affect areas of special environmental sensitivity (which are normally country-specific). From this list, those that are of great magnitude, extent or significance, or which involve areas of sensitivity, are selected for baseline studies. Though the scoping will normally result in the prediction of all potential impacts, the baseline studies that follow scoping may sometimes result in the identification of additional impacts.

The baseline studies, which provide the benchmark for the future, consist of recording the baseline levels of the environmental parameters that have been selected through scoping. This is normally done through the study of available documents, supplemented by field surveys. The socio-cultural and economic investigations should include the identification of affected communities and consultations with them.

Possible socio-economic changes and public health problems, if any, that may arise as a result of the farming project should be adequately investigated.

14.3.3   *Methods of impact identification*

Several methods or approaches have been proposed and used for identifying and evaluating environmental impacts. Out of approximately 100 methods, the United Nations Economic and Social Commission for Asia and the Pacific (ESCAP, 1985) selects the following eight classes as of importance:

- checklists,
- environmental evaluation,
- matrices,
- networks,
- overlays,
- environmental indices,
- cost–benefit analyses,
- simulation modelling workshops.

Descriptions of the important methods are contained in the ESCAP document and other literature listed in the references. Each method has its advantages and disadvantages, but the best-known are checklists, matrices and networks.

In its simplest form, a checklist is a list of environmental, social and economic factors that may be affected by the development. Such a checklist is only able to aid the identification of impacts, ensuring that no important impacts are overlooked. A checklist incorporating 'thresholds of concern', which expresses values that are desirable or that should not be reduced (Sassaman, 1981), could be of considerable help in evaluating environmental consequences. The elements, appropriate criteria and 'threshold of concern' values have to be selected specifically for each project. The main disadvantage of checklists is that they do not give any guidance on the ways an environmental component may be affected by one or more developmental features.

The matrix method serves to bring together all the project actions and the environmental components so as to identify the interactions. Developed for the US Geological Survey (Leopold *et al.*, 1971), the matrix is probably the one method that has been very widely employed, with amendments and additions to suit the needs of individual EIA studies. It displays a list of project activities against a list of environmental characteristics or impact indicators in order to identify cause-and-effect relationships. Column headings generally list the project activities, while the row headings show the environmental characteristics of the affected system. Each entry item on one set of headings can be related systematically to all items in the other set to ascertain whether an impact is likely.

The matrix can be used to measure and interpret impacts by describing them in terms of magnitude and importance on a common 1–10 or similar scale, where 1 is the least magnitude or importance and 10 is the greatest. Scores are assigned for magnitude and importance to all specific aspects that have been identified as representing a likely impact, and are included in the appropriate cell of the matrix. The cell is bisected by a diagonal line and the score for magnitude is placed in the top left-hand corner. The score for importance is placed in the bottom right-hand corner, as shown in Figure 14.4, which represents the EIA matrix of a project for the establishment of a 4200-ha fish farm in the Nile Delta described by Collinson (1980). Generally, the scores are accompanied by a plus or minus sign (+ or –) to indicate whether an impact is beneficial or adverse.

Unlike in the illustrated matrix, the score for importance is also given on the numerical scale in order to analyse the percentage number of impacts identified in each environmental system (natural, social and economic), as well as the overall percentage of impacts as shown by Braun (1990) in the EIA matrix of Mto-wa-Mbu irrigation and flood control project (Fig. 14.5). Here, the scoring system ranges from 1 to 5 (1 = very small impact, 2 = small impact, 3 = important impact, 4 = very important impact, and 5 = extremely important impact).

Besides the score for the magnitude and importance of the identified impacts, the cells also include the product of multiplying magnitude (M) with importance (I) as an assessment score of the impact in the upper right, accompanied by a plus or minus sign. The possible number of interactions of the environmental impact in the matrix is expressed in the number of activities in and out of the project (7), multiplied by the number of environmental factors (25). The possible number of interactions is 175 in the example, consisting of 77 in the natural system, 63 in the social system and 35 in the economic system, whereas the real interactions are 29, 23 and 14, respectively, of which 8, 14 and 14, respectively, are positive. From this, the overall percentage, of impacts generated by both actions in and out of the project for each of the three systems are calculated. Braun (1990) also showed that the degree of importance of the positive and negative impacts can be displayed in a visual way in a matrix using different tones or colour patterns.

Scoring system

- Magnitude: 1 – small; 2 – medium; 3 – large
- Importance: A – small; B – medium; C – large
- * interaction; either no impact or positive impact

Physical and chemical characteristics		Canals, drains and ponds	De-watering	Construction traffic	Material stock piling	Structures	Vegetation clearance	Groundwater	Surface water	Flow regulation	Water characteristics and alterations	Pumping	Pond, drainage/filling	Water quality	Water management	Vegetation development	Manure storage	Offsite traffic	Construction workers	Permanent staff (technical)	Permanent staff (manual)	Permanent staff (guardians)
Earth	Land form	3/B			2/A																	
	Soil constitution	3/A	2/*	1/A	2/A			1/A	2/A													
	Mineral resources	1																				
	Construction materials	3/A																				
Water	Surface	3/*	3/A		2/B	1/A			1/A			1/A	1/A	*								
	Groundwater	1/A	1/B		1/C	1/A		1/A	1/A	1/A		1/A	1/A	*								
	Recharge	1/A	1/A		1/A	1/A				1/A			1/A	*								
	Quality	1/A	3/*		1/C	1/A			1/A			1/A	1/A				1/C					
Air	Climate																					
	Quality				1/A	1/A	1/A										1/C					
Processes	Floods		*																			
	Erosion	*	*																			
	Silting		*																			

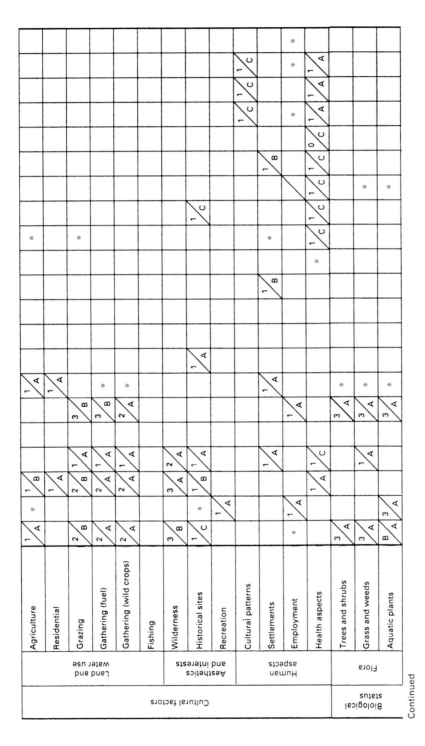

Figure 14.4 EIA matrix of a project for the establishment of a fish farm in the Nile Delta in Egypt. (From: Collinson, 1980)

Continued

Scoring system

Magnitude
1 – small
2 – medium
3 – large

Importance
A – small
B – medium
C – large

* interaction; either no impact or positive impact

	Construction activities for preparing the site						Water and land regime alterations				Operational activities							Labour requirements			
	Canals, drains and ponds	De-watering	Construction traffic	Material stock piling	Structures	Vegetation clearance	Groundwater	Surface water	Flow regulation	Water characteristics and alterations	Pumping	Pond, drainage/filling	Water quality	Water management	Vegetation development	Manure storage	Offsite traffic	Construction workers	Permanent staff (technical)	Permanent staff (manual)	Permanent staff (guardians)
Biological status — Fauna																					
Insects	3/A			*		1/A		*							*	2/B					
Aquatic animals	B/B	3/C																			
Birds	*	1/A	1/A	*		1/A		*				*			*	*					
Land animals	B/A	1/A	1/A	*		2/A		*							*						
Manmade works — Existing																					
Industries (salt extraction)		2/A					2/A	2/A	2/A		1/A	1/A		1/A							
Structures	1/C		1/A																		
Military	*		2/B																		
Manmade works — Proposed																					
New Ameriya City										1/A							1/A				
Recreation											2/B										
Free trade zone														2/C							
Others	1/A		1/A																		
Disease incident																					
Schistosomiasis	1/A	*						1/C	1/C	1/B		*	1/C	1/C	1/C			1/C	*	1/B	*
Malaria	1/A	*						1/C	1/C	1/B		*	1/C	1/C	1/C			1/C	*	*	
Others	1/A	*						1/C	1/C	1/B		*	1/C	1/C	1/C	1/C		1/C			

Figure 14.4 (continued)   EIA matrix of a project for the establishment of a fish farm in the Nile Delta in Egypt. (From: Collinson, 1980)

The major limitation of the matrix system is that it focuses on direct impacts between two items, and results in compartmentalizing the environment into separate items. The cumulative aspects of different impacts through different pathways are not considered. The network method is intended to respond to this need and deal with impact pathways and their implications for environmental features.

The network method was first proposed by Sorensen (1971, quoted in ESCAP, 1985) and involves the development of a stepped matrix of 'cause–condition–effect' to trace the nature of environmental interrelationships. The network takes the form of a relevance impact tree that relates and records secondary, tertiary and higher-order impacts. Its construction is based on answers to a series of questions relating to each item of the project activities, for example, what are the primary impact areas, what are the secondary impact areas, and what are the tertiary impacts within these areas.

Figure 14.6 is reproduced from ESCAP (1985) to show an example of a network. It represents the EIA network/stepped matrix of the Nong Pla Reservoir in the Mekong Basin. This approach allows the identification of both short-term and long-term effects, as well as direct and indirect impacts. It facilitates the environmental design or control as it traces the cause and effect of the relevant factors. However, the method does not evaluate the magnitude or significance of the impacts or make any explicit consideration of alternative projects or beneficial effects.

Probably because of the time it takes to develop the basic matrices required, this method does not appear to have been applied in many EIA studies.

14.3.4 *Prediction and evaluation*

Once the impacts have been identified by suitable methods, the next step would be to evaluate them and predict their extent and nature. For example, if change in the water quality is likely to have a significant impact on the environment of a fish farm project, it is necessary to predict the nature of changes in quality, such as the levels of suspended solids, BOD or nutrient concentration, and the extent of the changes brought on by the discharges. The baseline studies referred to earlier would provide information on the ambient levels. Based on this and the water discharge rates of similar fish farming operations, it should be possible to estimate a series of water quality conditions that may occur in the surrounding environment under different farming operations. If computer models are available, they can be used to calculate the water quality changes that may occur.

Predictions of environmental effects of development projects are based on physical, biological, socio-economic and anthropological data, and the quantification of these effects is often the most important and difficult part of EIA. Assessment and comparisons are made easier when impacts

ENVIRONMENT SYSTEM	ENVIRONMENT PARAMETERS	IN	PROJECT					OUT	+	-	Σ+	Σ-	Σ total
		Operation of irrigation system	Maintenance	Infrastructure works	Water distribution	Expansion of irrigation scheme	Agricultural production	Soil erosion					
SOCIAL	Tourism facilities	0	0	0	0	1 \ 2+ / 2	2 \ 4+ / 2	0	2	-	6	-	+6
	Social integration	0	0	0	0	5 \ 20+ / 4	4 \ 16+ / 4	0	2	-	36	-	+36
	In-migration problems	0	0	0	0	4 \ 16- / 4	4 \ 16- / 4	0	-	2	-	32	-32
	Maasai people	0	0	5 \ 25- / 5	0	5 \ 25- / 5	0	0	-	2	-	50	-50
	Human health	5 \ 20- / 4	1 \ 4+ / 4	0	5 \ 20- / 4	5 \ 20- / 4	0	0	1	3	4	60	-56
	Social conflicts	0	0	0	0	5 \ 20- / 4	4 \ 16- / 4	0	-	2	-	36	-36
	Employment growth	5 \ 20+ / 4	0	5 \ 20+ / 4	0	5 \ 20+ / 4	3 \ 15+ / 5	0	4	-	75	-	+75
	Increase urban services	0	0	0	2 \ 4+ / 2	0	1 \ 2+ / 2	0	2	-	6	-	+6
	Training activities	3 \ 9+ / 3	0	3 \ 9+ / 3	0	0	1 \ 3+ / 3	0	3	-	21	-	+21

ACTIONS

ENVIRONMENT PARAMETERS	IN			PROJECT			OUT					
	Operation of irrigation system	Maintenance	Infrastructure works	Water distribution	Expansion of irrigation scheme	Agricultural production	Soil erosion	+	−	Σ+	Σ−	Σ total
NATURAL												
Salinity and alkalinity of soil	4 | 12/3	0	0	4 | 12/3	0	0	0	−	2	−	24	−24
Soil erosion	0	0	0	0	3 | −12/4	0	5 | −25/5	−	2	−	37	−37
Wild life	0	0	3 | −15/5	0	5 | −25/5	0	5 | −25/5	−	3	−	65	−65
Ichthiofauna	0	2 | −4/2	5 | −10/2	0	0	0	0	−	2	−	14	−14
Water quality	0	2 | 4+/2	0	0	0	0	4 | −16/4	1	1	4	16	−12
Water sanitation	3 | −15/5	0	0	3 | −15/5	5 | −25/5	0	0	−	3	−	55	−55
Water flow	0	5 | 20+/4	0	0	2 | −8/4	0	5 | −20/4	1	2	20	28	−8
Water logging	3 | −9/3	4 | 12+/3	0	4 | 12/3	0	0	0	1	2	12	21	−9
Siltation problems	0	2 | 8+/4	0	0	0	0	5 | −20/4	1	1	8	20	−12
Flood control	2 | 8+/4	0	5 | 20+/4	0	0	0	2 | −8/4	2	1	28	8	+20
Drought control	4 | 16+/4	0	5 | 20+/4	5 | 20+/4	0	0	3 | −12/4	3	1	56	12	+44

ENVIRONMENTAL SYSTEM

Continued

Figure 14.5   EIA matrix of Mto-wa-Mbu irrigation and flood control project in Tanzania. (From: Braun, 1990)

System	Component	C1	C2	C3	C4	C5	C6	C7	C8	C9	C10	C11	C12	Total
ENVIRONMENTAL SYSTEM — ECONOMIC	Market of products	0	0	0	0	0	5 / 20+ / 4	5 / 20+ / 4	0	2	–	40	–	+40
	Increase production area	0	0	5 / 20+ / 4	0	0	5 / 20+ / 4	0	0	2	–	40	–	+40
	Increase crop production	3 / 15+ / 5	2 / 10+ / 5	5 / 25+ / 5	5 / 25+ / 5	3 / 15+ / 5	0	0	0	5	–	90	–	+90
	Income distribution	0	0	0	0	1 / 4+ / 4	5 / 20+ / 4	5 / 20+ / 4	0	2	–	24	–	+24
	Creation of growth pole	0	0	5 / 25 / 5	0	4 / 20+ / 5	4 / 20+ / 5	4 / 20+ / 5	0	3	–	65	–	+65
	+	5	7	6	3	8	8	–	37	–	–	–	–	–
	–	4	1	3	4	8	2	7	–	29	–	–	–	–
	Σ+	68	58	139	49	21	100	–	–	–	535	–	–	–
	Σ–	56	4	50	59	151	32	126	–	–	–	478	–	–
	Σ total	+12	+54	+89	–10	–30	+68	–126	–	–	–	–	–	+57

Figure 14.5 (continued) EIA matrix of Mto-wa-Mbu irrigation and flood control project in Tanzania. (From: Braun, 1990)

can be quantified. However, all impacts are not quantifiable. Even though methods have been proposed for measuring environmental intangibles (Coomber & Biswas, 1973), it is not easy to adopt them in all farming projects. As intangible consequences are important and cannot be ignored, an alternative is to include them in a qualitative form.

A qualitative assessment of an impact can be made on the basis of expert opinions, leading to a prediction of its magnitude. Ahmad & Sammy (1985) suggested the Delphi technique, which draws on the knowledge of a panel of experts on the subject being investigated, using individual assessment, statistical analysis and controlled feedback to arrive at a consensus.

The cost of impact quantification is a major consideration in EIA, and this can be optimized by determining the degree of accuracy required. During the scoping phase of the study, the degree of accuracy required can be set by a scrutiny of the nature of the impacts to determine whether the data gathered would require complex computerized models to derive the answers sought. It should be remembered that the quantified impacts are only predictions and not facts, and so there is bound to be some inherent uncertainty which cannot be completely avoided, but efforts have to be made to keep this uncertainty within acceptable limits.

In evaluating and describing environmental impacts, it is necessary to characterize them in terms of their spatial and time dimensions. Some impacts occur only in the immediate vicinity of the project, whereas others may affect a much wider area. Similarly, some impacts occur immediately, whereas others take a long time to become apparent. The environmental and social distributions of the impacts and the cumulative consequences of multiple impacts have also to be characterized. In most cases, the beneficial or adverse nature of impacts are obvious, but from a developmental point of view it is necessary to assess the reversibility of all the impacts.

It is often possible to reduce the adverse effects of development projects by suitable mitigation measures, even if their elimination proves impossible. Some of the mitigation measures that are feasible in aquaculture projects have been described in previous chapters. Each of these has associated costs. When applicable measures have been selected, their costs have to be calculated and the level of environmental impact requantified, taking into account the beneficial effect of the mitigation measure. The acceptability of the proposed measures has to be determined by adequate consultations with the affected communities and relevant interest groups. Depending on the mitigation measures, more than one project alternative may emerge from such computations.

Comparison of alternatives is the essence of the assessment, where all available technical information is synthesized and a full description of each project alternative is prepared by combining the environmental losses and gains with the economic costs and benefits. For the purpose of comparing alternatives, the positive and negative environmental

PROJECT ELEMENT	CAUSAL FACTOR				
Water resources development	Dam and reservoir / Irrigation system			Condition	
Altered element	Development phase	Initial	Changes		Final
PHYSICAL RESOURCE					
Surface water quality	★	more water storage	more nutrient enrichment		disturbed aquatic habitat
		less water flow	more salinity		disturbed coastal zone characteristics
	☆	more phosphate	more productivity		
		more pesticides & fertilizer utilization	more residual pollution		more toxic accumulation in food chain
Ground water hydrology		—	—		—
Ground water quality		—	—		—
Soils	□	flooded area	– – – –		loss of agriculture
		intensive land use	soil loss		loss of soil fertility
Geology/Seismology		—	—		—
Erosion Sedimentation	□	more sedimentation trapping	less storage capacity dam		less dam life
		more bank erosion	more turbidity		less water quality
Climate	□	changed relative humidity	changed microclimate		changed rainfall
ECOLOGICAL RESOURCE					
Fisheries	□	more productivity	more job opportunity		more income
	☆	less fish migration	less fish population		less income
Aquatic biology	□	less riverine habitat	less species in reservoir		less species diversity
		less nutrient	less primary productivity		less aquatic population
Terrestrial wildlife		—	—		—
Forest	□	less of deciduous,	forest		change in climate

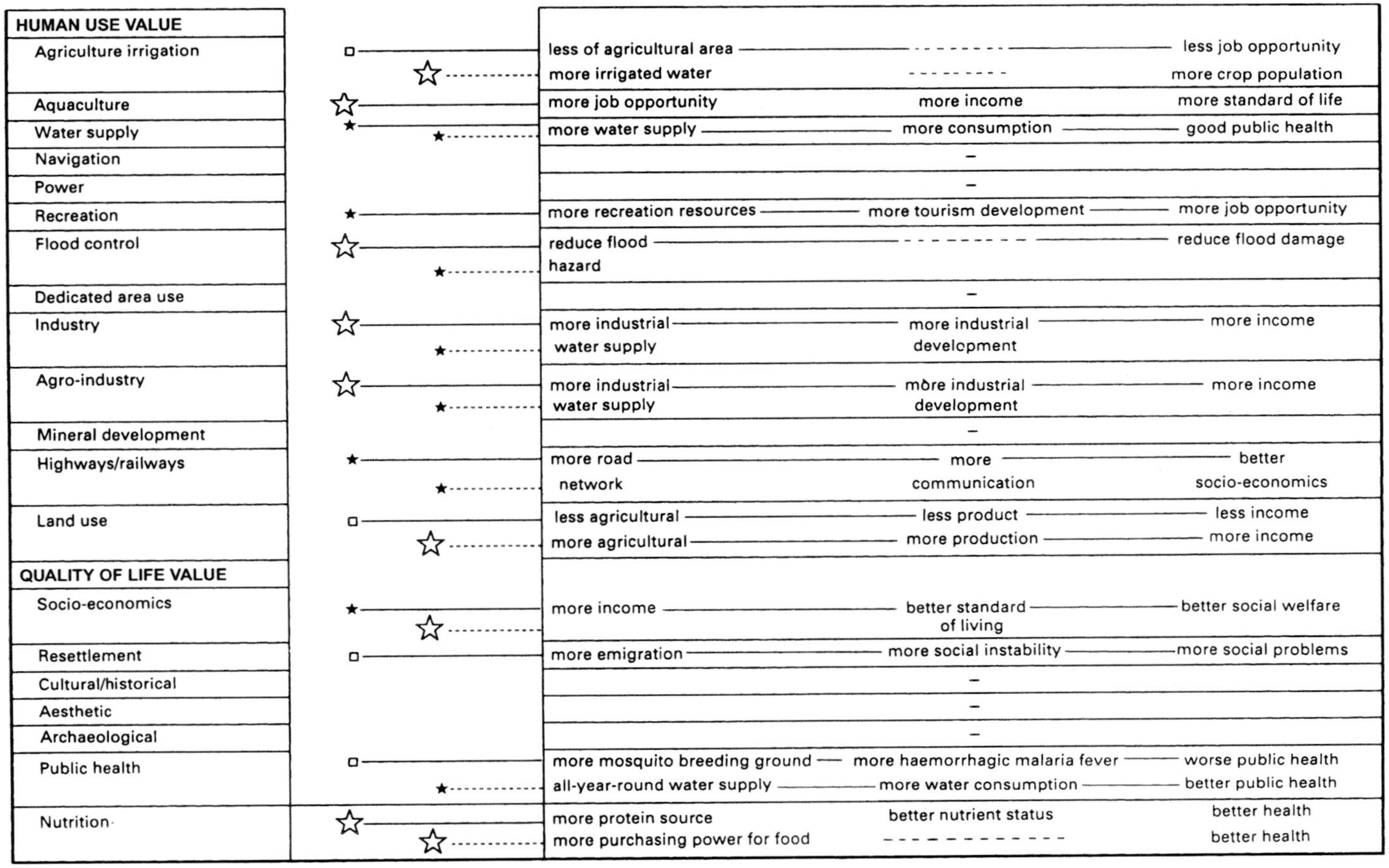

Figure 14.6   Network/stepped matrix for Nong-Pla Reservoir in the Mekong Basin. (From: ESCAP, 1985)

impacts and the economic costs and benefits are used. The economic data may be obtained through EIA or from a parallel economic analysis. Cost–benefit analysis is the simplest means of comparing alternatives from both environmental and economic aspects. This will require environmental impacts to be converted into economic equivalents and expressed as costs or benefits. Recommendations can then be made on the basis of cost–benefit analysis of each alternative. This should simplify decision-making.

The problem is that some impacts cannot be reduced to cash equivalents. Though procedures have been proposed for placing economic values on intangibles, they continue to be controversial.

Other solutions suggested range from simple ranking of alternatives to graphical and importance-weighting techniques. The objective is to arrive at a preference ranking of alternatives on the basis of economics, and a ranking of alternatives on the basis of environmental impacts, using methods such as the Leopold *et al.* (1971) matrix, or checklist format. Other methods suggested include pair-wise comparisons, which consist of comparing two options at a time, and repeating this until each option has been compared with every other option, assigning scores on the basis of preference. The preferred option is the one that receives the highest score. In the graphical approach, the relative economic and environmental desirability of the options are plotted on a graph and the final selection is made on the basis of this display; for example, the project costs in dollars are plotted against environmental ranking on a scale of 1 to 10. If the graph does not show the best choice but, for example, shows two options as being better than the others, it will be necessary to select either environmental or economic considerations as the deciding factor.

Another approach is the use of weighted ranking, which involves giving 'importance weights' to economic and environmental consequences. The options are ranked in terms of economic and environmental acceptability. The ratios of importance of economic to environmental concerns are determined in advance. The weighted ranks are the product of the actual rank and the 'importance weight'. If the ranking system awards rank 1 to the best environmental or economic options, the overall preferred option will be the one with the lowest total score.

14.3.5   *EIA document and decision-making*

The product of EIA is the document that contains the results and recommendations of the study, supported by all the relevant reference material. Its main purpose is to aid decision-making on development projects and to serve as a management tool, especially to ensure that potential problems are foreseen and addressed in the project design and operation. Since decision-makers may not always be people who have been closely associated with the studies, it is necessary to take special care in the preparation of the document to answer the questions that are likely to be asked and present the results of the studies in an easily understandable and usable manner. An executive summary of the main findings is an essential part of the document.

In countries with enabling legislation on EIA, the decision-making mechanism and the competent authority would have been designated and appropriate procedures defined. Dispute settlement procedures may also have been set up in case objections are raised about the decisions made by the competent authority. Appropriate government agencies, concerned members of the public and experts in relevant disciplines, and interested non-governmental organizations, may be given an opportunity to comment on EIAs relating to projects or enterprises that have major environmental significance.

In making decisions, the information and recommendations contained in the EIA document are examined impartially. The decision-makers also consider political realities, the perceptions of the local population and the relationship of the proposed project to others in the area. The competent authority may, therefore, refer the document for expert or public review, or ask the developer to undertake additional studies on questions that have not been adequately answered, and submit a revised EIA document.

If the project is finally approved, the decision-maker will be required to decide on a plan for reducing conflicts, if any, about the project, which may include compensation for affected groups and public education, and to designate institutional responsibilities for overseeing the environmental management of the project and its monitoring.

14.3.6 *Monitoring*

As EIA is essentially a prediction exercise, it is necessary to verify the validity of the predictions made in the absence of certainty and to determine the effectiveness of the mitigation measures adopted. Until recently, this was generally carried out by the environmental authority through a system of monitoring and post-audit. It was considered as an isolated post-project action to increase knowledge for the purpose of improving EIA of future projects.

The importance of impact monitoring in relation to EIA, as well as project implementation, has now been increasingly recognized, and this activity, referred to as EIA audits or post-project analysis (PPA), is recommended to be carried out as a part of the EIA. It involves environmental studies undertaken during the implementation of a given activity to ensure or facilitate its implementation in accordance with the terms imposed by the environmental assessment process, or to learn from the particular activity being studied. It provides valuable feedback to project implementation and helps to identify unanticipated impacts for immediate corrective measures (GESAMP, 1996a).

Such a study is of special importance in aquaculture projects in view of the scarcity of verified knowledge on environmental impacts of different culture systems under various operational conditions.

PPA or impact auditing is orientated to determining (1) whether the EIA document includes the complete range of impacts that have actually occurred, and (2) whether the individual predictions made in the document are accurate. It is done on the basis of the data obtained from impact monitoring, and may lead to re-evaluation of the earlier assessments and consequent changes in project management, besides providing useful information for improving future EIAs of similar projects.

Rapid changes taking place in aquaculture technologies often make it possible to introduce adaptations to impact mitigation procedures, and improve farming practices. So monitoring and impact auditing during project implementation can be of considerable value in enhancing its efficiency.

In some types of project, such as engineering systems where there are possibilities of hazardous events or catastrophic accidents occurring, techniques of risk analysis may be employed to cope with the uncertainties inherent in predictive assessments.

Chapter 15
Mitigation of Adverse Effects

Most development activities give rise to some form of adverse effects, some of which may be of major significance. Planning the mitigation of such effects starts with the impact identification or scoping stage of the EIA exercise, and is evaluated by comparing options along with their economic costs and benefits. Identification of mitigation measures would normally be based on experience gained elsewhere in similar projects, although novel ideas and procedures can also be considered. The main mitigatory practices employed in aquaculture are summarized here.

15.1 Land and Water Use

The main types of extensive land areas used for land-based aquaculture are wetlands, including marshes and mangrove swamps, and agricultural lands. The changes brought about by conversion into pond farms are not irreversible, unlike some of the other competing uses, although conversion back to their original state will entail expense and time.

The traditional practice of reclaiming swamps for agricultural lands through an intermediate stage of aquaculture (Pillay, 1958), the recent practice of converting unproductive agricultural land into aquaculture farms, and the re-afforestation of mangroves show that the changes can be successfully reversed. The combined culture of fish or crayfish and agricultural crops such as rice is also practised in some areas. Similarly, pen and cage culture in open-water areas can be phased out, if considered necessary, to restore ecological conditions similar to those that existed before aquaculture was started in the area. However, aquaculture projects are presently conceived as long-term sustainable developments, and so this alternative is seldom considered. It then becomes necessary to undertake mitigatory measures to reduce environmental impacts.

Mangroves are now accepted as a resource subject to multiple use, and the strategy for their management will naturally be based on national developmental priorities, including the preservation of the existing ecosystem. Social and economic considerations have an over-riding role and these alone will require the allocation of appropriate areas for aquaculture, selected and managed with minimum impact on the ecological functioning of the mangrove system as a whole.

The introduction of intensive farming systems has been advocated to reduce areas under culture. This would involve the use of pumping to irrigate the ponds, which will also facilitate the location of farms in back-swamps outside the limits of the normal tidal regime, thus reducing the utilization of the more productive zones of mangroves. It will have the added advantage of reducing the need for excavation of mangrove soil for dike construction, and thereby reducing the production of acid sulphate soil and the leaching of acid water into the environment (Cook *et al.*, 1984).

The need to conserve the tall mangrove stands along the margins of swamps where new accretion of detritus occurs has been described in Chapter 5. It is generally only the secondary growths that are cleared for farm construction. It is a general practice in many areas to plant mangroves on the dikes of fish farms, which will partially compensate for the loss of natural growth and also stabilize the pond dikes.

Thorhaug (1987) has reviewed the methods and results of seagrass and mangrove restoration efforts in the Caribbean and to some extent in Asia (mainly of mangroves). Aquaculture has not been found to be detrimental to seagrasses, except that there is the likelihood of some destruction when farm dikes are built through seagrass beds. Planting and maintenance are much more difficult in tidal areas than in terrestrial environments because of high tidal amplitudes, currents, wave action and the occurrence of cyclones and typhoons. However, as a result of work for over two decades on mangroves and over a decade on seagrasses, practical methods are now available for their cultivation.

The most successful method of planting seagrass such as *Zostera* is the use of plugs of the grass. These consist of blades, roots, rhizomes and sediments extracted from natural beds and planted into holes made on the site for restoration, and anchored in turbulent areas by cement collars, chicken wire or low barriers to reduce wave energy. Planting of seeds or seedlings has also proved cost-effective when suitable anchors are used to stabilize them. Shoots and turfs have been planted successfully, using various anchoring devices.

Though available information on the recovery of animal communities is scanty, documented observations seem to indicate substantial colonization in restored seagrass beds. The cost–benefit analysis of seagrass restoration is made difficult by the fact that the benefits are indirect and mainly to fisheries, and their value as food of biologically or economically important fishery species has not been quantified. Thorhaug (1987) cites the cost of experimental restoration as from US$1500 to US$20 000 ha^{-1}, which according to her contrasts favourably with the value of US$150 000 ha^{-1} placed on estuaries by UNEP in 1980.

A common practice is to turf the dikes of pond farms in coastal areas with salt-resistant grasses and creepers, which normally thrive in such areas, mainly to reinforce the dikes and walkways and to prevent soil erosion. However, this also contributes to the preservation of the characteristic species in salt marshes.

Mangrove restoration has been undertaken in several countries of south-east and west Africa and many of the more important species are now under cultivation. Successful re-afforestations have been undertaken in large mangrove forest areas by government forest departments, private industry and community groups, although their value in terms of animal repopulation or contribution to commercial fisheries has not been determined.

The most successful results have been obtained when young trees (30–100 cm in height) are planted, rather than seeds, propagules or seedlings. Survival rates are higher in low-energy areas. High-energy areas require larger trees and provision of energy dampeners or barriers. It is reported that proper elevation within the inter-tidal zone is necessary if the seaward trees are to survive (Thorhaug, 1987).

The cost of mangrove restoration varies considerably within and between countries, and the species of mangrove and technique of planting adopted. Lewis (1982) gives the actual cost of planting *Rhizophora mangle* in a commercial project (0.91 m spacing) with purchased propagules at US$6545 ha^{-1} and with collected propagules at US$12 500 ha^{-1}. The estimated cost of planting 6-month-old purchased seedlings of *R. mangle*, *Avicinnia germinans* and *Laguncularia racemosa* in the above spacing works out to between $2510 and $12 103 ha^{-1}. It is reported that rapid recolonization occurs soon after the leaf canopy has developed. No published quantification of the benefits is available.

The advantages of introducing intensive farming techniques to reduce the extent of mangrove or marshy areas under culture have been mentioned in Section 5.3. This is an alternative that deserves consideration, as experimental work has shown that a pumped water supply for such farms is feasible and economically viable (Gedney *et al.*, 1984). However, waste production will simultaneously increase with intensification, and additional land and costs will be involved in the treatment and disposal of the wastes (see Section 5.3.2).

The main source of water in land-based freshwater aquaculture is surface water abstracted from rivers, streams, lakes and irrigation systems. Pumped groundwater is often preferred for hatchery purposes and may, in a smaller number of cases, be used in production farms as well. The use of public water resources for aquaculture may be regulated but usually there are very few problems with the use of surface water for aquaculture, except in arid regions. Use of irrigation water may be severely restricted, particularly during dry seasons, but there are no major constraints for less intensive uses such as cage culture in irrigation reservoirs and canal systems.

The quantity of water consumed in aquaculture depends on the culture system and the stock density in the farm. Tanks and race-way systems normally require large quantities of water but a good percentage is discharged back into ground or surface waters. In the case of pond farms under stagnant and semi-stagnant conditions, loss of water by seepage and evaporation can be high, depending on climatic conditions.

Seepage varies to a considerable extent, depending on soil conditions, area of pond surface and dike construction. Boyd (1981) observed an average seepage of $1.3-2.5$ cm d^{-1} in experimental ponds in Auburn (USA). Loss due to evaporation was 0.48 cm d^{-1} (Parson, 1949). Besides the surface run-off, the ponds received direct rainfall and, in some cases, groundwater inflows to compensate for the loss, and the excess inflow is discharged through spillways to maintain the required water-level. In sub-tropical countries, evaporative losses will be much lower, around 0.5 cm d^{-1} (Hepher and Pruginin, 1981). The average water loss in Europe is estimated to be between 0.4 and 0.8 cm d^{-1} (see Section 6.1).

In Israel, water requirements for ponds, allowing for an estimated loss of $1-2$ cm d^{-1}, may work out to between 35 000 and 60 000 m^3 ha^{-1} yr^{-1} in order to maintain an average water depth of 1.5 m throughout the growing season of about 8 months. When such large quantities of water have to be bought, even intensive systems of aquaculture cease to be viable and the ponds have to be converted for multiple uses such as irrigation-cum-aquaculture. Rice-cum-fish culture is another option for economic use of water and multiple use of resources.

Though water use in terms of production may be low in intensive aquaculture, in absolute terms high rates of water use are characteristic of such systems, as the water quality has to be maintained at a high level even when aeration and other means of improving quality stand-ards are employed. There are some exceptions to this, as in pond culture of some air-breathing Asian catfish that can survive in very limited quantities of stagnant water loaded with organic matter (see Fig. 15.1).

Abstraction of large quantities of water from groundwater sources has been found to cause environmental problems, salination of agricultural land and land subsidence, as mentioned earlier. Water abstraction from surface waters can also produce adverse consequences, even if they are comparatively less pronounced. These may include reduction in flow rates, increased siltation, and alterations in biological communities and behavioural patterns of fish populations in the inflow sources. It is essential, therefore, to prevent indiscriminate abstraction of ground or surface water, even though the observed problems are localized and not very widespread.

Available information from analogous conditions can be used for preliminary estimates of water requirements, as well as availability, but it is necessary to validate them by local investigations and analyses. Loss of water by seepage can be reduced by proper site selection and careful construction of ponds. The reduction of water flow in streams or rivers (which serve as the source of inflow and reception of discharges) can be minimized by keeping the intake and outflow of farms as close together as possible.

Some unique attempts to develop fish farming in national parks in combination with bird sanctuaries and protected wildlife, as in Hortabagh in Hungary, have not been economically successful, evidently because of the high rate of predation by some of the birds and mammals. It is

Figure 15.1 An intensive catfish pond in Thailand being drained. Note the remaining pond water containing large quantities of organic matter.

preferable to locate farms as far away as possible from their habitats or areas where they congregate, not only to reduce chances of predation in the farms but also to prevent disturbance to the birds and protected mammals, especially in their breeding areas.

Protective devices can be used to prevent predation of aquaculture stocks. Electric fences or wires powered by car batteries have been used successfully to deter otters and mink from fish farms. Unplanned trapping of birds and predatory mammals, often as a result of faulty use of antipredation nets in cage farms, can easily be reduced by care on the part of the farmer. As mentioned in Chapter 11, shooting is not a particularly efficient means of controlling bird or mammal predation. It has been suggested, therefore, that shooting of at least the protected species should be strictly controlled and monitored, with a view to enhancing or maintaining their populations at the required levels.

As the use of foreshore areas, bays and fjords creates the maximum conflicts and environmental concerns, special care and consideration of public interests are necessary in aquaculture development in these areas. As far as possible, large cage farms should be located in isolated areas, away from shore fronts of urban areas and scenic spots frequented by local inhabitants or tourists. While unsightly structures and installations can be avoided, it may not be feasible to conceal floating cages entirely. Nevertheless, efforts can be made to screen them from major viewing points by trees, fencing or ground modelling. Visibility of the farms from viewing points can be reduced by choosing dark, subdued and non-reflective colours for the floating parts of the farm, except for buoys and floats which need to be easily visible to aid navigation. Landscaping of land-based structures and buildings should be attempted. The additional costs and inconvenience caused by these measures may be more than compensated by the positive response of local communities at the planning and operational stages of the farms.

Of equal importance in site selection, farm design and operation is the potential for transmission of water-borne diseases. The precautions described in Chapter 5 have to be strictly followed to avoid the colonization of intermediate hosts of these disease-causing organisms in aquaculture farms.

15.2   Culture Practices

In selecting culture systems and practices, aquaculturists seldom give adequate weighting to waste production and disposal, as compared with economic yields of the cultured species. Waste reduction is an integral part of good husbandry, and disregard of this can be counter-productive as, in the long run, the sustainability of the operations will be affected by the degradation of the external environment. It is, therefore, important to estimate, in as precise a manner as feasible, the quality and quantity of the waste produced and discharged from the farm, and its fate and effect in the receiving waters. The available information has been reviewed briefly in Chapters 7 and 10.

The main forms of wastes to be considered are suspended solids and dissolved nutrients, especially nitrogen and phosphorus. As described in Chapter 7, the major sources of these wastes are uneaten or spilled feeds and faecal matter. In tidal ponds, the inflows may contain considerable quantities of organic matter. This, as well as unutilized primary production resulting from fertilization, may add to the loads of solid and dissolved wastes in effluents from these ponds. The need to avoid over-fertilizing of ponds through excessive application of organic or chemical fertilizers is more widely recognized, as the results, such as algal blooms, are clearly visible.

Over-feeding is not so apparent, especially when automatic feeding techniques are used. It has been estimated that feed losses (when processed feeds are used) can vary from 5 to 20% and over-feeding can reduce feed digestibility and increase faecal production significantly. Experienced farmers know that hand-feeding is the most rational way of feeding, as the feeding can be directly observed and losses minimized (see Fig. 15.2). In areas where labour costs prohibit the adoption of hand-feeding, at least regular monitoring of feed consumption can be practised. The use of computerized feeding systems based on automatic monitoring of the environment and food conversion ratios, practised in many north European salmon farms, is reported to be very effective in minimizing feed losses.

Figure 15.2 Hand-feeding of salmon in a cage farm.

The physical characteristics of feeds are of major importance in increasing the pollution load of farm effluents. From this point of view, dry pellets are far superior to wet or moist pellets, and trash fish diets have to be avoided. Palatability and high water-stability of the pellets would improve consumption of feeds and reduce losses. Unstable pellets break down easily into fine particles and interfere with uniform settlement in sedimentation ponds. The sinking rate of pellets is important in salmonid pond farms, as well as in floating cages, to allow the fish to capture the feed before it sinks to the bottom. Feed wastage can be minimized by the use of low-density extruded pellets.

The soluble contents of effluents are influenced by the composition of the feed. As indicated in Section 3.4, phosphorus and nitrogen are of major concern from the environmental viewpoint. In order to minimize the waste of phosphorus, its concentration in the feeds should be kept low and its bio-availability and feed conversion improved. Recent investigations of salmonid nutrition (e.g. Wiesmann *et al.*, 1988) indicate that current levels of phosphorus in fish feeds are in excess of requirements. For example, the average concentration in trout diets is about 15.9 g P kg^{-1} as against a requirement of 4.5 g P kg^{-1}, and 11.7 g P kg^{-1} as against a requirement of 3 g P kg^{-1}.

The form of phosphorus in the feed affects its availability. Some forms, such as phytin phosphorus from soya bean meal and phosphorus from wheat, are poorly converted and result in higher waste loads (Ketola, 1982). Use of high-quality fish meals and oils results in low phosphorus output levels per unit fish weight (Crampton, 1987). Another option is to substitute defluorinated rock phosphate, which is less polluting and dissolves only slightly in water.

Even though the nitrogen budget of fish and other cultivated organisms is not fully understood, it is clear that the amount excreted depends on feeding rate, digestibility and availability. Beamish and Thomas (1984) estimate that on average about 60% of dietary nitrogen is excreted in rainbow trout. The assimilation of protein depends very much on the quality of the ingredients used in the preparation of feed. If the amino-acid composition of the feed protein and its bio-availability do not meet the requirements of the cultured animal, there will be considerable protein wastage, occurring in farm effluents as nitrogenous waste compounds such as ammonia. Digestibility of nitrogen varies directly with dietary protein and lipid concentration, whereas faecal nitrogen varies inversely with dietary protein and is not dependent on lipid concentration.

The use of poor-quality carbohydrates in feed manufacture can result in increased solids and waste of BOD in farm waters. While excess carbohydrates in effluents increases BOD and causes sewage fungus, insufficiency results in an increase of ammonia and other nitrogenous wastes. Though oils and fats used in high-energy diets 'spare' protein and, therefore, reduce ammonia excretion, they contribute to the development of visible surface scums and increased BOD in farm effluents.

It is evident from the above that feed quality and conversion ratios have a very considerable influence on the characteristics of the effluents, and even with the present knowledge, it is possible to improve the quality of feeds and manufacture 'low pollution' feeds. Reduction in phosphorus contents, control of dietary nitrogen in relation to metabolism, and improvement in physical characteristics of feeds along with optimum feed usage can lead to a substantial reduction in nutrient loadings of effluents.

15.3   Waste Treatment

The handling and disposal of wastes, especially solid wastes, are easier in land-based pond farms than in shellfish and cage farms in open waters. At present there are no proven techniques of removing or treating wastes from shellfish or cage farms. The only practical procedure is to disperse sedimented wastes periodically by means of submersible pumps or similar devices. In smaller inland water areas, this procedure may not be advisable because of the limited area normally available for spreading the wastes. If large accumulations have to be dispersed, caged fish or shellfish rafts have to be moved out of the site to prevent mortality as a result of low dissolved oxygen concentration and the release of hydrogen sulphide and methane.

The possibility of rotating sites to allow time for mineralization of wastes has been referred to earlier in this chapter. Japanese shellfish farmers leave farmed areas fallow for about 18 months after harvests, and during this period repeatedly trawl the area to oxygenate and mineralize sediments. Atlantic salmon cage farms in Scotland are reported to observe fallowing for periods ranging from 4 to 51 weeks (University of Stirling *et al.*, 1990), during which time the nutrient wastes become diluted and the sediments recover. The single mooring method of cage installations is another means of dispersing solid wastes, albeit only within a limited area. Experimental work on collecting solid wastes from floating cages into funnel-shaped collectors, and pumping the sludge away at intervals, has been mentioned in Chapter 7. However, none of these has been developed to a stage for practical application so far.

Waste reduction and treatment are now regularly practised in many land-based farms. In stagnant and semi-stagnant pond farms of Asia, effluent discharges take place usually after harvesting. After the long retention period of about one year or more, much of the solids would have settled at the pond bottom. The sediments are then removed for use as fertilizer for land crops or dried and treated to serve as fertilizer in the pond itself when the pond is prepared for the next crop. If the final draining of the pond is delayed by a few days after harvest, the sediments resuspended during fishing would have resettled, and the solid and dissolved nutrient load of the discharge will not be very significant.

In intensive pond farms that maintain a flow-through, the time of retention of water passing through the farm is much less. In Europe it is reported to range from 20 to 720 minutes, depending on the water flow (Alabaster, 1982). In such cases it will be valuable to incorporate a suitable settlement treatment system to reduce the solid load in the effluents, and thereby reduce a significant proportion of the BOD and the phosphorus bound into the solids.

There are a number of ways of treating effluents from pond farms. Sand filtration, microstraining and air flotation have been tried, but simple sedimentation has proved to be more cost-efficient in commercial farms. The main constraints in sedimentation are the high volume of effluents involved and the low concentration of solids in them compared to domestic waste waters. The process of settling involves the falling out of suspension of solids having densities higher than that of water, when the water flow is reduced under gravitational force. In commercial pond culture, settling or sedimentation ponds or basins are commonly used. For a pond with a water use of $1200 \text{ m}^3 \text{ h}^{-1}$, for example, a relatively large area of 500 m^2 will be required for best results (Mantle, 1982).

The effectiveness of sedimentation ponds depends on the design, the surface area available for settling, and the flow or retention time of the effluent. A rectangular shape is generally preferred, as a steady laminar flow is ensured by keeping the width of the inlet and outlet weirs the same as that of the pond (see Fig. 5.6). According to Arceivale (1983), a length–width ratio higher than 4 will improve settling efficiency, and Mangelson and Watters (1972) found that installation of baffles would increase the mean residence time of a pond by as much as 75% by enhancing the length–width ratio of the flow path. While the retention time is undoubtedly an important factor, the actual time required appears to vary considerably.

Particulate size of the solids is also important, as smaller and less dense particles require lower current velocities than larger and denser particles. It is necessary, therefore, to reduce fragmentation of particles through turbulence. Liao (1970*b*) found a period in excess of 3 days necessary to remove over 60% of the suspended solids, but others have reported retention times of less than a few hours as adequate.

From a study of 16 settlement ponds in commercial fish farms in the UK, Henderson and Bromage (1988) concluded that settling efficiency is greatly reduced when concentration in the influent is <10 mg l^{-1}, but concentration can be increased by pretreatment. Settlement could be optimized by maintaining a low fluid velocity of less than 4 m min^{-1} in order to minimize turbulent resuspension. According to the University of Stirling *et al.* (1990), recent studies have shown that up to 90% of suspended solids, 60% of BOD and 50% of total phosphorus loads can be removed by settlement treatment, although the system performance is highly variable.

Mechanical devices for effluent treatment are also in use in hatcheries and production units on a more limited scale. The simplest procedure is to screen them by means of a stationary filtering mesh to remove particulate matter. Clogging of the filter and high maintenance costs are major drawbacks to this method. A self-cleaning or rotating filter has been developed to reduce this problem but the rotating action of the screen can cause higher concentrations of dissolved nutrients.

Another type of self-cleaning filter known by the trade name 'Triangel filter' is claimed by the manufacturer to remove 90% of suspended solids, 80% of total phosphorus, 70% BOD and 40% total nitrogen. This is achieved by the relatively quick separation of solids from the effluent water, thus reducing the time available for leaching of soluble material from solid wastes, as well as reducing the level of dissolved wastes (especially ammonia and dissolved reactive phosphorus) that can occur in settling ponds. In actual use, the removal of solids and BOD has been reported to be lower than in the settlement ponds, but the dissolved nutrients in the filtered effluent were found to be lower.

The self-cleaning filter, as well as the swirl concentrator (see Chapter 5) described by Warrer-Hansen (1982), are reported to be operated in conjunction with sedimentation tanks in the areas where they are in operation.

Sludge collected in the settlement ponds should be cleared at least once a year, or at shorter intervals if the accumulation obstructs water

Figure 15.3 Polyculture of seaweeds, scallops and abalone on the Chinese coast, which reduces hypernutrification of the environment.

flow in the ponds. The best use of the sludge is as fertilizer for agricultural crops. In rural areas, the sludge can be digested in septic tanks for production of biogas.

Though a bio-filtration technique can be employed to prevent accumulation of ammonia and nitrite in recirculating and experimental systems, its application in commercial aquaculture, even in tropical countries with higher temperatures, is problematic due to the large volumes of effluent involved and the low concentrations.

Aquatic macrophytes such as water hyacinth and duckweeds have been grown in effluent-receiving waters for the extraction of nutrients, but the multiplication and dispersal of such plants, which are difficult to control even with repeated use of herbicides, make their use risky. Seaweed culture seems to be an efficient means of removing high concentrations of nutrients in the environment.

The polyculture of *Laminaria* and other seaweeds, along with scallops and abalone in coastal areas, has been reported to be very effective in controlling hypernutrification in the coastal areas of north-east China (see Fig. 15.3). It has been so efficient that additional fertilizers have to be sprayed along the coast to encourage rapid growth of the seaweeds and of the planktonic algae that form the food of the shellfish. Folke and Kautsky (1989) have suggested that eutrophication as a result of cage farming can be avoided by integrating it with mussel farming, which leads to nutrient depletion.

The species combination in polyculture of Chinese and Indian carp in pond farms generally includes at least one scavenging species that feeds on faecal matter of the other species. This serves to reduce the load of faecal matter and thereby of soluble nutrients. The whole concept of polyculture based on full utilization of fish pond resources is geared to the reduction of wastes, and is therefore environmentally very sound indeed.

Among other aquaculture practices that have potential environmental consequences are the use of chemicals and antibiotics. When the existing guidelines are followed and permitted non-persistent chemicals are used, there is very little chance of adverse impacts on the environment. Problems arise when the necessary precautions are ignored and harmful chemicals are allowed into the environment. The prophylactic use of antibiotics is ineffective but has the potential to cause resistance among bacterial populations or even transmit antibiotic-resistance to other bacteria. It can thus become dangerous to human health. Therefore, its use in aquaculture has to be discouraged.

Proper sanitary measures have to be observed to prevent transmission of diseases between farms. Eggs or juveniles should be bought from disease-free hatcheries, and when in doubt, should be quarantined for an appropriate period to ensure that they are disease-free. If proper precautions are not observed, new diseases can be introduced into an area through uncontrolled import of exotic species. Guidelines are available (see Chapter 9) for controlling such introductions, but require legislative support to ensure adequate compliance, as discussed in Section 16.2. Similarly, the adoption of depuration techniques for ensuring the safety of aquaculture products for human consumption would have to be controlled by legislative measures.

As described in Chapter 12, shellfish poisoning caused by occasional toxic algal blooms has been responsible for several deaths and disabilities. As there are no known means of preventing them, the establishment of appropriate monitoring systems in areas prone to algal blooms or red tides is essential to provide early warning, so that necessary precautionary measures can be taken to prevent the consumption of contaminated products and to avoid economic losses to the farmers.

Even though there is as yet no conclusive evidence, the potential for behavioural interactions between escaped fish from farms and wild populations has to be considered. It is feared that such interactions may adversely affect gene pools through the introduction of non-adaptive genotypes into wild populations. Therefore, a precautionary approach is warranted, and all possible precautions have to be taken to prevent the escape of farmed fish, and particularly hybrids and other genetically engineered stocks, into the environment.

Chapter 16
Research and Regulation

16.1 Modelling the Environment

The inadequacy of available information has been repeatedly emphasized in discussions on most aspects of environmental impacts of aquaculture. This is due to the fact that aquaculture research institutions, with some notable exceptions, have devoted very little attention to environmental problems outside the farm limits.

It is only the emergence of large marine commercial farms and consequent widespread public concern that resulted in the increasing restrictions and opposition in many areas that triggered environmental research relating to aquaculture. Most of such research is still centred in industrialized countries, largely because of the cost of long-term investigations involved and the multi-disciplinary expertise required for studying the wide spectrum of factors that govern the environmental consequences of aquaculture.

Because of the urgent need for action in connection with development programmes, temporary solutions are sought by interpolating information from what are considered analogous cases. This is only justifiable if there are enough analogous cases covering the wide variety of climatic, hydrographic, biological, technological and socio-economic conditions in which different aquaculture systems are presently practised. Obviously, at present, the tentative use of deducible knowledge of this nature is justifiable only in certain areas and in respect of certain culture systems. However, even in such cases decisions or conclusions made on the basis of this will need to be reviewed and modified, if necessary, when specific information becomes available. There is, therefore, an urgent need to organize interdisciplinary research to determine the environmental impact of the more important aquaculture systems under typical environmental conditions.

The elements of such a research programme are too diverse to be enumerated here, but the more important ones have been identified in earlier chapters. The main aim of the research should be to generate information to predict, with a reasonable degree of accuracy, the environmental impact of aquaculture systems, and to develop cost-effective methods for eliminating or ameliorating the adverse impacts.

Predictive modelling can be an effective tool in making development decisions. A number of efforts have been made to model the risk of

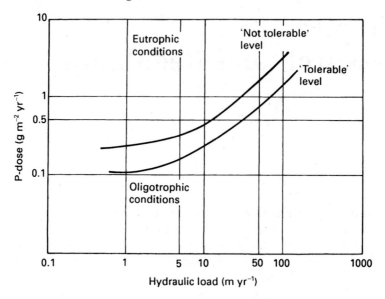

Figure 16.1 Vollenweider (1968) model for assessing the relationship between trophic state and phosphorus loading in a lake.

eutrophication in lakes and freshwater lochs. The Vollenweider models (1968; 1975) were intended to assess the effect of phosphorus loading on the trophic state of lakes, as illustrated in Figure 16.1, based on the mass balance equation. The response to phosphorus (P) loading depends on the relationship between annual phosphorus and hydraulic loads. Phosphorus loading is expressed as loading per unit area (mg P m^{-2} yr^{-1}). The hydraulic load is the run-off from the catchment area per unit lake area (m yr^{-1}).

As nitrogen is often the limiting nutrient for bioproduction in the marine environment, Håkanson *et al.* (1988) consider the model unsuitable for marine areas. It has been used in impact assessments of freshwater cage farms in Sweden and Norway (Carlsson, 1988). The need for precise information on phosphorus loadings from sources other than the cage farm, and the assumption of a steady trophic state before the installation of the cages, are the major problems in the use of this model. Such information has to be gathered by field measurements or estimated from data on catchment area land use (see Reckhow & Chapra, 1983).

The model widely used for predicting impacts of fish farming on the concentration of phosphorus in the water column is that of Dillon and Rigler (1974). The model is based on the simple mass–balance equation

$$P = \frac{T_w \times L \times (1 - R)}{z}$$

where P = predicted phosphorus concentration (mg m^{-3})
   T_w = water residence time (years)
   L = areal loading rate (mg P m^{-2} yr^{-1})
   R = sedimentation coefficient
   z = mean depth (m)

The sedimentation coefficient estimates the loss of phosphorus to the sediments, and the water flow determines the amount of phosphorus lost through the outflow. When applied to predict the effect of phosphorus loading by a cage farm in a freshwater lake or loch, the estimated increase in phosphorus as a result of cage farming is added to the background concentration to give the final predicted concentration. This model also has the same problem in determining the background phosphorus concentration, which is complicated by seasonal variations in both external and internal sources of phosphorus, and losses from the water column through flushing and sedimentation. This problem might be overcome by measuring phosphorus levels at the time of spring overturn (in temperate climates) before the spring stratification develops.

The water residence time, which is the time taken for one exchange of the total volume, is calculated by dividing the total volume of the water body by the inflow. Inflow can be estimated from calculations of the catchment area and effective rainfall (i.e. total rainfall minus evapotranspiration).

Areal loading rate is a measure of the phosphorus load per unit surface area, which can be calculated from the annual fish production in tonnes multiplied by the estimated phosphorus discharged for each tonne of fish production, which is about 10 kg.

Phosphorus can be lost within a water body by sedimentation, when the internal concentrations reach equilibrium with the external loadings (Reckhow & Chapra, 1983), and so an estimate of the sedimentation loss is included in the model. The most commonly used equation to calculate this loss is the one described in OECD (1982):

$$R = 0.426 \exp(-0.271\, q_a) + 0.574 \exp(-0.00949\, q_a)$$

where
   R = retention coefficient
   q_a = hydraulic load z/T_w

Mean depth is estimated by field survey or from hydrographic tables.

The main limitation in the use of this model is the assumption of a steady-state balance between phosphorus loads and water flow, which seldom occurs because of seasonal and annual variations in fish production, external and internal inputs and flow. It is reported (University of Stirling *et al.*, 1990) that the sensitivity of the model has recently been improved by taking these variations into account, and closer correlations between predicted and measured concentrations have been achieved. Even with further refinements, the basic problem of measuring all the potential nutrient inputs into the water body remains, and this makes accurate modelling difficult. Water movements within the water body affect concentrations of phosphorus in the water column, and incomplete mixing and thermal stratification make it necessary to monitor dissolved subtances from a variety of depths to obtain a representative profile. In the absence of such data, the values assigned to parameters of a mass balance model may be erroneous.

The problems mentioned above highlight the lack of some of the basic information needed for predicting the impact of aquaculture with any degree of accuracy, even in confined areas such as freshwater lakes, lochs and similar water bodies. It highlights the need to apply existing predictive models with care and to follow up predictions with appropriate monitoring in order to modify farming operations as required.

Other models of value in predicting impacts of aquaculture are hydrological models, especially of water quantity and quality in fresh waters (NERC, 1975; Carr, 1989), and modelling of sediment impacts in marine environments (Weston & Gowen, 1988; Håkanson *et al.*, 1988). Attempts have also been made to model the carrying capacity of tropical inland waters used in cage fish culture (Beveridge, 1984).

Although there are many empirical and mechanistic models for predicting the input of organic matter from the sea-bed, quantitive connections between input of organic matter and ecological changes have not yet been developed (GESAMP, 1996b). Various guidelines have been suggested for estimating maximum rates of input of organic matter on theoretical grounds, taking into account factors such as rates of dispersion, resuspension and microbial decomposition. Environmental capacity, in other words, assimilative capacity, which measures the resilience of the natural environment in the face of stock enhancement or aquaculture development, is of importance for estimating the rate at which nutrients can be added without triggering eutrophication. Aure and Stigebrandt (1990) attempted to model capacity in terms of level of input of organic wastes in fjords of Norway used for cage culture, based on published data, on the assumption that the site was depositional (rather than erosional). The climatic conditions can cause considerable differences and the guidelines suggested are unlikely to apply directly to tropical areas. The capacity of the sediments to absorb organic matter may be 3–4 times greater in the warm water of the tropics. In spite of this, the model can be used as a means of predicting effects of nutrient enrichment from inputs of organic matter. This can be extended to other systems (GESAMP, 2001) if estimates can include vertical mixing and horizontal exchange of water. It is not easy to translate site-specific models to new locations, despite attempts to make models generic (Grant and Bacher, 1998).

Models have been developed for the evaluation of intangible elements of the environment, such as its attractiveness and aesthetic qualities.

Attractivity models are based on the demand for an environmental experience as measured in user participation or satisfaction (Coomber & Biswas, 1973; Shafer *et al.*, 1969; Leopold, 1969). However, these are not known to have been used or adapted for use in decision-making with regard to the siting or design of aquaculture facilities.

16.2 Regulatory Measures

It is logical to believe that the environmental management measures described earlier would be implemented voluntarily by all concerned, as they are directed to the sustainable development of an important and promising food production industry, without degrading the environment, but the long-term nature of some of the benefits and, more importantly, the restraints and restrictions that these measures involve, make it necessary to have appropriate regulatory mechanisms based on statutory authority (Howarth, 1999). Many countries already have rules and regulations that are applied to aquaculture, although only few are specifically meant to regulate or protect aquaculture. As observed by Van Houtte *et al.* (1989):

> the aquaculturist must cope with a complex network of laws and regulations which deal with land tenure, water use, environment protection, pollution prevention, public health and fisheries in general.

This results in confusion, conflicts and overlapping of provisions, and causes frustration to the industry. At the same time, countries with specific sets of rules on aquaculture in the Western industrialized world are perceived to have erred too much towards caution and have tended to over-regulate, causing more or less the same amount of frustration to the industry.

As the needs and conditions in different countries vary so widely, and are likely to change in response to economic developments, a single model aquaculture law cannot be formulated for the use of all countries. It follows that laws and rules should be framed on a national basis to meet the needs of aquaculture development and environmental protection under conditions existing or likely to develop in each country. However, the need for a unified national aquaculture legislation that would promote sustainable development, protecting and enhancing the environment, cannot be over-emphasized.

The special importance of the location of farm sites and the hydrological features of the adjacent water bodies in preventing pollution and degradation of the environment have been emphasized earlier. Zoning and allocation of sites for aquaculture not only allows development in appropriate areas, but also helps in avoiding conflicts with other users of such areas while promoting the optimum use of land and water resources. While zoning has to be based on socio-economic, technical and market considerations, the actual allocation of sites would require the intervention of the appropriate regulatory authority based on national or provincial legislation. In this sense, zoning and allocation of sites does not seem to have been attempted so far, though some countries have regulations to protect certain areas under different frameworks.

Van Houtte *et al.* (1989) cite instances of (1) the Mexican *Ecological Balance and Environment Protection Law* (1988) established under the *Federal Hunting and Fishing Law*, which provides for areas where the exploration, exploitation or use of natural resources is regulated, and (2) the Presidential Decree of 1978 in the Philippines, where certain areas and projects (which include dikes for fish farm development) have been declared environmentally critical. Here, development activities require an Environmental Compliance Certificate issued by the President or his representative on the basis of an environmental impact statement.

The fisheries law of the Republic of Korea has established 'protective areas' which *inter alia* designate areas 'for seeding of marine animals and plants or the growing of fish larvae'. The Ministerial Agreement of 1978 in Ecuador allows the establishment of shrimp hatcheries along the Ecuadorian coast only in technically specified areas decided by a commission of the governmental authorities concerned. To protect agricultural land from salination, a security, strip of 200 m is enforced between fish ponds and hatcheries under a regulation of 1975. Besides these, some south-east Asian countries have declared the maximum areas of mangroves that can be cleared for aquaculture, thereby regulating access to aquaculture sites in mangrove swamp areas.

Directives for marine fish farming have tried to limit the locations for new farms. Scottish ministers believe that further development of marine finfish farming developments should be limited on the east and north coasts (Scottish Executive, 1999).

As discussed earlier, properly organized environmental impact assessment could be an invaluable aid in planning and implementing large-scale aquaculture enterprises. Though the benefits of such an assessment should be clear to entrepreneurs, the time, cost and expertise that it involves can become disincentives, and persuade them to take avoidable risks. To prevent this, it will be necessary to insist on appropriate environmental impact assessments and incorporation of the required mitigation measures in the project from its very early stages.

Some countries already have regulations that require impact studies to secure a licence for aquaculture, but many of them need to update the guidelines for the assessments using available additional knowledge. Small-scale aquaculture, which is unlikely to have any significant impact on the environment, is generally exempt from the need for such assessment, except in especially sensitive areas. Usually, the scale is determined on the basis of annual production, which according to present practice seems to vary, but appears to be around 100 t. Decisions on this have to be based on local conditions, bearing in mind that a large number of small-scale farms located in proximity along a watercourse can have as much or even greater impact than a single large-scale farm.

The enforcement of rules and regulations to benefit aquaculture without degrading the environment and, if possible, enhancing the environment, would require the establishment of appropriate mechanisms supported by knowledgeable field and management personnel.

Environmental research relating to aquaculture should be accorded the importance it deserves and all regulatory measures should be reviewed and modified as additional information becomes available. Even if the present volume of aquaculture production and its contribution to gross national product may not match the public investment involved, the potential of this emerging industry would more than justify concerted efforts to direct its growth on a sustainable basis.

References & Further Reading

Ackefors H. (1999) Environmental impact of different farming technologies. In *Sustainable Aquaculture—Food for the Future?* (Ed. by N. Svennevig, H. Reinertsen & M. New) 145–69. A.A. Balkema, Rotterdam.

Ackefors H. & Enell M. (1994) The release of nutrients and organic matter from aquaculture systems in Nordic countries. *Journal of Applied Ichthyology* 10, 225–41.

Ackefors H. & Södergren A. (1985) Swedish experience of impact of aquaculture on the environment. ICES/CM. E:40/Sess.W:1–7.

ADB/NACA (1998) *Aquaculture Sustainability and the Environment.* Report on a Regional Study and Workshop, RETA, 5534.

Ahmad Y.J. & Sammy G.K. (1985) *Guidelines to Environmental Impact Assessment in Developing Countries.* Hodder and Stoughton, London, 52p.

Alabaster J.S. (1982) Report of the EIFAC Workshop on Fish Farm Effluents. Silkeborg, Denmark, 26–28 May 1981. *EIFAC Tech. Pap.*, 41, 166p.

Alabaster J.S. & Lloyd R. (1980) *Water Quality Criteria for Freshwater Fish.* Butterworths for FAO, London, 279p.

Alderman D.J. *et al.* (1984) Crayfish plague in Britain. *J. Fish. Dis.*, 7, 401–5.

Aldridge C.H. (1988) Atlantic salmon pen strategies in Scotland. *Proc. Aquacult. Int. Congr.* (Vancouver), 28p.

Alfsen C. (1987). Shellfish culture in France. SEAFDEC, Bangkok, 96p.

Alikunhi K.H. (1957) Fish culture in India. *Farm Bull.*, 20, 144p.

Allen D.A., Austin B. & Colwell R.R. (1983) Numerical taxonomy of bacterial isolates associated with a freshwater fishery. *J. Gen. Microbial.*, 29(7), 2043–62.

Allen G.H., Busch R.A. & Morton A.W. (1979) Preliminary bacteriological studies on wastewater-fertilized marine fish ponds, Humboldt Bay, California. In *Advances in Aquaculture* (Ed. by T.V.R. Pillay & W.A. Dill). Fishing News Books, Oxford, England, 492–8.

Allen G.H. & Hepher B. (1969) Recycling of wastes through aquaculture and constraints to wider application. In *Advances in Aquaculture* (Ed. by T.V.R. Pillay & W.A. Dill). Fishing News Books, Oxford, England, 478–87.

Allen K.R. (1949) The New Zealand grayling, a vanishing species. *Tuatura*, 2, 22–7.

Allendorf F.W. & Phelps S.R. (1980) Loss of genetic variation in a hatchery stock of cutthroat trout. *Trans. Amer. Fish Soc.*, 109, 537–43.

Alongi D.M. (1987) The influence of mangrove-derived tannins on intertidal meiobenthos in tropical estuaries. *Oceanologia* (Berlin), 71, 537–40.

Alzieu C. & Héral M. (1984) Ecotoxicological effects of organotin compounds on oyster culture. *Ecotoxicological Testing for the Marine Environment*, 2, 187–95.

Alzieu C. & Portmann J.E. (1984) The effect of tributyl on the culture of *C. gigas* and other species. *Proc. Annual Shellfish Conf.*, 15, 1–15.

Anderson D.M. (1989) Toxic algal blooms and red tides: a global perspective. In *Red Tides: Biology, Environmental Science and Toxicology* (Ed. by T. Okaichi, D.M. Anderson & T. Nemoto). Elsevier, New York, 11–16.

Anon. (1973) Position of American Fisheries Society on introduction of exotic aquatic species. *Trans. Amer. Fish. Soc.*, 102, 274–6.

Anon. (1986) *Tambak or mangrove*. Paper presented in the Panel Discussion on Mangrove Greenbelt, Ciloto, Bogor, Indonesia (mimeo).

Arakawa K.Y. (1973) Aspects of eutrophication in Hiroshima Bay viewed from transition of cultured oyster production and succession of marine biotic communities. *Nihon Kaiyo Gakkai-Shi*, **11**(2), 43–8 (in Japanese).

Arakawa K.Y., Kusuki Y. & Kamigaki M. (1971) Studies on biodeposition in oyster beds. 1. Economic density for oyster culture. *Venus*, **30**(3), 113–28 (in Japanese).

Arceivale S.J. (1983) Hydraulic modelling for waste stabilisation ponds. *J. Environ. Eng.*, AICE, **109**(5), 265–8.

Aston R.J. (1981) The availability and quality of power station cooling water for aquaculture. In *Aquaculture in Heated Effluents and Recirculation Systems, Vol. 1*. (Ed. by C. Tiews). Schriften der Bundesforschungsanstalt für Fischerei, Hamburg. 39–58.

Aure J. & Stigebrandt A. (1990) Quantitative estimates of the eutrophication effects of fish farming in fjords. *Aquaculture* **90**, 135–56.

Austin B. (1982) Taxonomy of bacteria isolated from a coastal marine fish rearing unit. *J. Appl. Bact.*, **53**, 253–68.

Austin B. (1983) Bacterial microflora associated with a coastal marine fish rearing unit. *J. Mar. Biol. Assoc. U.K.*, **63**, 585–92.

Austin B. (1985) Antibiotic pollution from fish farms: effects on aquatic microflora. *Microbial. Sci.* **2**(4), 113–7.

Austin B. & Allen-Austin D. (1985) Microbial quality of water in intensive fish rearing. *J. Appl. Bact.* Symposium Supplement **14**, 59, 2075–265.

Avnilmech Y. (1998) Minimal discharges from intensive fish ponds. *World Aquaculture* **29**, 32–7.

Ayres P.A. & Cullum C.C. (1978) Paralytic shellfish poisoning. An account of investigations into mussel toxicity in England, 1968–77. *Fisheries Research Report, Lowestoft*, **40**, 1–23.

Babbit H.E. & Baumann E.R. (1958) *Sewerage and sewage treatment*. John Wiley & Sons, Inc., New York.

Baklien Å. (1989) Floating raceway. *World Aquaculture* **20**(1), 75–6.

Baluyut E. (1983) A review of inland water capture fisheries in Southeast Asia with special reference to fish stocking. *FAO Fish. Rep.*, **288**, 13–57.

Barbier E.B. (1987) The concept of sustainable economic development. *Environ. Conserv.* **14**, 101–10.

Barg V.C. (1992) *Guidelines for the Promotion of Environmental Management of Coastal Aquaculture Development*. FAO Fisheries Technical Paper No. 328. FAO, Rome.

Barthelmes D. (1984) On the problem of the intensive carp rearing in lakes and eutrophication. *Acta Hydrochim. Hydrobiol.* **12**(2), 153–61 (in German).

Baticados M.C.L. *et al.* (1990) *Disease of penaeid shrimp in the Philippines. Aquaculture Extension Manual No. 16*. SEAFDEC Aquaculture Department, Tigbauan, Iloilo, Philippines.

Beamish F.W.H. & Thomas E. (1984) Effects of dietary protein and lipid on nitrogen losses in rainbow trout, *Salmo gairdneri*. *Aquaculture*, **41**, 359–71.

Bergheim A. *et al.* (1984) Estimated pollution loadings from Norwegian fish farms. II. Investigations 1980–1987. *Aquaculture*, **28**, 347–61.

Bergheim A. & Selmer-Olsen A.R. (1978) River pollution from a large trout farm in Norway. *Aquaculture*, **14**, 267–70.

Bergheim A. & Siversten A. (1981) Oxygen consuming properties of effluents from fish farms. *Aquaculture*, **22**, 185–7.

Bergheim A., Siversten A. & Selmer-Olsen A.R. (1982) Estimated pollution loadings from Norwegian fish farms. I. Investigations 1978–79. *Aquaculture*, **28**(3–4), 347–61.

Beveridge M.C.M. (1984) Cage and pen fish farming: carrying capacity models and environmental impact. *FAO Fish. Tech. Pap.*, **255**, 131p.

Beynon J.L. *et al.* (1981) Nocturnal activity of birds on shrimp mariculture ponds. *J. World Maricul. Soc.*, **12**(2), 63–70.

Bird C.J. & Wright J.L.C. (1989) The shellfish toxin domoic acid. *World Aquaculture*, **20**(1), 40–1.

Bisset R. (1989) *Introduction to EIA methods.* Presented at 10th International Seminar on Environmental Impact Assessment and Management, University of Aberdeen, Scotland, UK.

Bjorklund H.J., Bonndestam G. & Bylund G. (1990) Residues of oxytetracycline in wild fish and sediments from fish farms. *Aquaculture*, **86**, 359–67.

Blogoslawski W.J. (1988) Ozone depuration of bivalves containing PSP: Pitfalls and possibilities. *J. Shellfish Res.*, **7**, 702–5.

Blogoslawski W.J. & Stewart M.E. (1978) Paralytic shellfish poison in *Spisula solidissima*: Anatomical location and ozone detoxification. *Marine Biology*, **45**, 261–4.

Bodvin T. *et al.* (1996) Clean technology in aquaculture—production without waste products? *Hydrobiologia* **327**, 83–6.

Bonn E.W. & Follis B.J. (1967) Effects of hydrogen sulfide on channel catfish, *Ictalurus punctatus*. *Trans. Amer. Fish. Soc.*, **96**, 31–6.

Bourne N. (1979) Pacific oysters, *Crassostrea gigas* (Thunberg) in British Columbia and the South Pacific Islands. In *Exotic Species in Mariculture* (Ed. by R. Mann). MIT Press, Cambridge, USA, 1–53.

Boyd A.J. & Heaseman K.G. (1998) Shellfish mariculture in the Benguela System: water-flow patterns within a mussel farm in Saldanha Bay, South Africa. *Journal of Shellfish Research* **17**, 25–32.

Boyd C.E. (1978) Effluents from catfish ponds during fish harvest. *J. Environ. Qual.* **7**, 59–62.

Boyd C.E. (1981) *Water quality in warmwater fish ponds.* Agricultural Experiment Station, Auburn University, Auburn, 359p.

Boyd C.E. (1985) Chemical budget for channel catfish ponds. *Trans. Amer. Fish. Soc.*, **114**, 291–8.

Boyd C.E. & Massaut L. (1998) Soils in pond aquaculture. *Aquaculture Asia*, **3**(1), 6–7.

Braaten B., Ervik A. & Bofe E. (1983) Pollution problems on Norwegian fish farms. *Aquaculture Ireland*, **16**, 6–10.

Braun R. (1990) *Environmental impact assessment of Mto-wa-Mbu, Tanzania, Irrigation and Flood Control Project.* ILO, Geneva, 101p.

Bromage N., Henderson J.P. & Watret R. (1989) How to design a settlement pond. *Fish Farmer*, **12**(3), 41.

Brown J.H. (1989) Antibiotics: their use and abuse in aquaculture. *World Aquaculture*, **20**(2), 34–5, 38–9, 42–3.

Brown J.R., Gowen R.J. & McLusky D.S. (1987) The effect of salmon farming on the benthos of a Scottish sea loch. *J. Exp. Mar. Biol. Ecol.*, **109**(1), 39–51.

Brunies A. (1971) Taint of mineral oils in mussels. *Archiv. für Lebensmittel*, **22**, 63–4.

Buras N. *et al.* (1987) Microbial aspects of fish grown in treated wastewater. *Water Research*, **21**(1), 1–10.

Butcher A.D. (1967) *A changing aquatic fauna in a changing environment.* IUCN Publications, New Series, **9**, 197–218.

Butz I. & Vens-Cappell B. (1982) Organic load from the metabolic products of rainbow trout fed with dry food. In *Report of the EIFAC Workshop on Fish Farm Effluents* (Ed. by J.S. Alabaster). *EIFAC Tech. Pap.* **41**, 73–82.

Bye V.J. & Lincoln R.F. (1986) Commercial methods for the control of sexual maturation of rainbow trout (*Salmo gairdneri* R.). *Aquaculture*, **57**(1–4), 299–309.

Cadwallader P.L. (1978) Some causes of the decline in range and abundance of native fish in the Murray–Darling river system. *Proc. Royal Soc. Victoria*, **90**, 211–24.

Caldwell R.S. (1975) *Hydrogen sulfide effects on selected larval and adult marine invertebrates*. Water Resource Research Institute, **31**, 27p.

Canter L. (1977) *Environmental impact assessment*. McGraw-Hill, New York.

Canzonier W.J. (1998) Public health component of bivalve shellfish production and marketing. *Journal of Shellfish Research* **7**, 261–66.

Cardwell R.D. & Sheldon A.W. (1986) A risk assessment concerning the fate and effects of tributyltins in the aquatic environment. *Oceans '86 Conference Record, Organotin Symposium, Vol. 4*, 1117–29.

Carlsson S.A. (1988) Report on suspended solids from land-based fish farms in Sweden. In *National contributions on suspended solids from land-based fish farms*. Papers presented at the First Session of the EIFAC Working Party on Fish Farm Effluents (Ed. by M. Pursiainen). The Hague, 29–30 May and 1 June 1987, 74–8.

Carpenter R.L. *et al.* (1974) The evaluation of microbial pathogens in sewage and sewage-grown fish. *Environ. Protect. Tech. Ser.*, Washington, DC. (EPA-660/2-74-041), 46–55.

Carr O.J. (1989) *Fish farm pollution model instruction manual*. Environmental Advisory Unit, Liverpool University, Liverpool.

Carss D.N. (1988) *Piscivorous birds at Scottish fish farms*. Proceedings of Fish Farming and Conservation Conference, Argyll Bird Club, Oban, Sept. 1987.

Carss D.N. (1990) Concentrations of wild and escaped fishes immediately adjacent to fish farm cages. *Aquaculture*, **90**(1), 29.

Chandler J.R. (1970) A biological approach to water quality management. *J. Water Pollut. Control*, **69**(4), 415–22.

Chen P.H. & Hsu S.T. (1986) PCB poisoning from toxic rice-bran oil in Taiwan. In *PCB and the Environment, Vol. III* (Ed. by J.S. Waid). CRC Press, Boca Raton, Florida, 22–37.

Chia L.S. (Ed.) (1988) *Environmental management in Southeast Asia, directions and current status*. Faculty of Science, National University of Singapore, Singapore, 211p.

Chiang H.C. & Lee J.C. (1986) Study of treatment and reuse of aquacultural wastewater in Taiwan. *Aquacult. Eng.*, **5**, 301–12.

Chiba K. (1981) Present status of flow-through and recirculation systems and their limitations in Japan. In *Aquaculture in Heated Water Effluents and Recirculation Systems, Vol. 2* (Ed. by K. Tiews). Schriften der Bundesforschungsanstalt für Fischerei, Hamburg, 343–55.

Cholik F. & Poernono A. (1987) *Development of aquaculture in mangrove areas and its relationship to the mangrove ecosystem*. In *Papers contributed to the Workshop on Strategies for the Management of Fisheries and Aquaculture in Mangrove Ecosystems*, Bangkok, Thailand, June, 1986 (Ed. by R.H. Mepham). *FAO Fish. Rep.*, (**370**) Suppl., 93–104.

Choluteca Forum (1996) *Choluteca Declaration*, October, 1996, *World Aquaculture* **28**, 38–9.

Chua T.E., Paw J.N. & Guarin F.Y. (1989a) The environmental impact of aquaculture and the effects of pollution on coastal aquaculture development in Southeast Asia. *Mar. Pollut. Bull.*, **20**(7), 335–43.

Chua T.E., Paw J.N. & Tech E. (1989b) Coastal aquaculture development in ASEAN: the need for planning and environmental management. In *Coastal Area Management in Southeast Asia: Policies, Management, Strategies and Case Studies* (Ed. by T.E. Chua and D. Pauly). *ICLARM Conf. Proc.*, **19**, 57–70.

Clareboudt M.R. *et al.* (1994) Fouling development and its effect on the growth of juvenile giant scallops (*Placopecten magellanicus*). *Aquaculture* **121**, 327–42.

Cleary J.J. & Stebbing A.R.D. (1985) Organotin and total tin in coastal waters of southeast England. *Mar. Pollut. Bull.*, **16**, 350–5.

Coleman M.S. *et al.* (1974) Aquaculture as a means to achieve effluent standards. *Environ. Protect. Tech. Ser.*, Washington, D.C. (EPA-660/2-74-041), 199–214.

Coleman D.J. (1977) Environmental impact assessment methodologies: A critical review. In *Environmental Impact in Canada: Approaches and Processes.* (Ed. by M. Plewes & J.B.R. Whitney). Institute for Environmental Studies, University of Toronto, 35–59.

Collinson R.I. (1980) Environmental impact assessment—in theory and in practice. In *The Environmental Impact of Man's Use of Water, Part I* (Ed. by S.H. Jenkins). *Water Sci. Technol.*, **13**(6), 105–14.

Commission of the European Communities (2002) Commission recommendation of 4 March 2002 on the reduction of the presence of dioxins, furans and PCBs in feeding stuffs and foodstuffs. *Official Journal of the European Communities*, **9**(3) L 67/69.

Conte F. (1984) Economic impact of paralytic shellfish poison on the oyster industry in the Pacific United States. *Aquaculture*, **39**, 331–43.

Cook H.L., Pongswana U. & Wechasitt S. (1984) Recommendations for construction and management of brackishwater aquaculture ponds in areas with acid sulfate soil. Field Document, Coastal Aquaculture Demonstration and Training Project, FI:DP/MAL/77/008, No. 2, 243–60.

Coomber N.H. & Biswas A.K. (1973) *Evaluation of environmental intangibles.* Geneva Press, Bronxville, New York, 75p.

Cooper P.F. & Hobson J.A. (1989) Sewage treatment of reed bed systems: the present situation in the United Kingdom. In *Constructed Wetlands for Wastewater Treatment* (Ed. by D.A. Hammer). Chelsea, Lewis, 774–9.

Cope O.B. (1964) Sport fishery investigations. In *The Effects of Pesticides on Fish and Wildlife.* U.S. Fish. Wild. Ser. Circ., **226**, 51–63.

Costa-Pierce B.A. & Roem C.M. (1989) Waste production and efficiency of feed use in floating net cages in eutrophic tropical reservoirs. In *Reservoir Fisheries and Aquaculture Development for Resettlement in Indonesia* (Ed. by O. Soemarwoto & B.A. Costa-Pierce). ICLARM Tech. Reports 23.

Crampton W. (1987) How to control phosphorus levels. *Fish Farmer*, July/Aug. 1987, 38–9.

Crown Estate (1987) *Fish Farming: Guidelines on siting and design of marine fish farms in Scotland.* Scotland Crown Estate Guidelines, 13p.

Cruz de la A. (1979) The functions of mangrove. In *Proceedings of the Symposium on Mangrove and Estuarine Vegetation in Southeast Asia.* BIOTROP Spec. Publ., (**10**), 125–38.

D'Abramo L.R.D. & Hargreaves J.A. (1997) Shrimp aquaculture at the crossroads: pathways to sustainability. *World Aquaculture* **28**, 27–39.

Dahl E. & Yndestad M. (1985) Diarrhetic shellfish poisoning (DSP) in Norway in the autumn 1984 related to the occurrence of *Dinophysis* spp. In *Toxic Dinoflagellates* (Ed. by D.M. Anderson, A.W. White & D.G. Daden). Elsevier, New York, 495–500.

Dahlbäck B. & Gunnarsson L.Å.H. (1981) Sedimentation and sulfate reduction under a mussel culture. *Mar. Biol.*, **63**(3), 269–75.

Davies F.R.E. *et al.* (1958) Shellfish toxin in cultivated oysters. *Canadian J. Public Health*, **49**, 286–7.

D'Croz L. & Kwiecinski B. (1980) Contribución de los manglares a las pesquerias de la Bahia de Panama. *Rev. Biol. Trop.*, **28**(1), 13–29.

Devoe M.R. *et al.* (1985) Coastal wetland impoundments in South Carolina: Policy issues and research. *Estuaries*, **8**(2B), 58A.

Dillon P.J. & Rigler R.H. (1974) A test of a simple nutrient budget model predicting the phosphorus concentration in lake water. *J. Fish. Res. Bd. Canada*, **31**, 1771–8.

Dixon J.A. (1989) Valuation of mangroves. *Tropical Coastal Area Management*, **4**(3), 1 and 3–6.

Donaldson E.M. (1997) The role of biotechnology in sustainable aquaculture. In *Sustainable Aquaculture* (Ed. by J.E. Bardach) 101–26. John Wiley & Sons Inc., New York.

Doumenge F. (1989) L'aquaculture in Equateur. In *Aquaculture, Vol. 2*, (Co-ord. G. Barnabe). Technique et Documentation—Lavoisier, Paris, 1229–42.

Downing K.M. & Merkens J.C. (1955) The influence of dissolved oxygen concentrations on the toxicity of un-ionized ammonia to rainbow trout (*Salmo gairdneri* Richardson). *Ann. Appl. Biol.*, **43**, 243–6.

Doyle J. *et al.* (1984) The impact of blooms on mariculture in Ireland. ICES Special Meeting on the Causes, Dynamics and Effects of Exceptional Marine Blooms and Related Events. Copenhagen, Oct. 1984.

Dugdale R.C. (1967) Nutrient limitation in the sea: dynamics, identification and significance. *Limnol. Oceanogr.*, **12**, 685–95.

ECE (1990) *Post-project analysis in environmental impact assessment.* ECE/ENVWA/11, Economic Commission for Europe, UN, New York, 54p.

ECHCP (2000) Assessment of dietary intake of dioxins and related PCBs by the population of EU Member States. *European Commission Health & Consumer Protection Directorate-General. Directorate C—Scientific Opinions.* Report of Experts Participating in Task 3.2.5, 7 June 2000, Brussels.

Edwards A. & Edelsten D.J. (1976) Marine fish cages—the physical environment. *Proc. Roy. Soc. Edinburgh (B)*, **75**, 207–21.

Edwards P., Polprasert C. & Wee K.L. (1987) Resource recovery and health aspects of sanitation. *AIT Research Report*, **205**, 324p.

Egidius E. & Moster B. (1987) Effect of NEGUVON and NUVAN treatment on crab (*Cancer pagurus, C. maenas*), lobster (*Homarus gammarus*) and blue mussel (*Mytilus edulis*). *Aquaculture*, **60**, 165–8.

EIFAC (European Inland Fisheries Commission) (1973) Water quality criteria for European freshwater fish. Report on ammonia and inland fisheries. *Water Res.*, **7**, 1011–22.

EIFAC (1988) Report of the EIFAC Working Party on Prevention and Control of Bird Predation in Aquaculture and Fisheries Operations. *EIFAC Tech. Pap.*, **51**, 79p.

Eknath (1993)

Ellis J.E., Tackett D.L. & Carter R.R. (1978) Discharge of solids from fish ponds. *Progr. Fish-Cult.*, **40**, 165–6.

Enell M. (1987) Environmental impact of cage fish farming with special reference to phosphorus and nitrogen loadings. ICES Council Meeting (collected papers), 13p.

Enell M. (1995) Environmental impact of nutrients from Nordic fish farming. *Water Science Technology*, 31: 61–71.

Enell M. & Löf J. (1983) Environmental impact of aquaculture—sedimentation and nutrient loadings from fish cage culture. *Vatten*, **39**(4), 364–75 (in Swedish).

Enell M., Löf J. & Bjorklund T.C. (1984) *Fiskkasseodling med Rening, Teknisk beskrivning och reningseffekt.* Institute of Limnology, Lund University, Sweden. (LUNBDS/NbII-3069)/1–34 (1984).

Environmental Protection Agency (EPA-USA) (1974*a*) *Aquaculture projects—Requirements for approval of discharges.* Federal Register 39(115), Part II, 1974, 20770–5.

Environmental Protection Agency (EPA-USA) (1974*b*) *Development document for proposed effluent limitation guidelines and new source performance standards for fish hatcheries and farms.* National Field Investigation Center, Colorado, 237p.

ESCAP (1985) *Environmental impact assessment—Guidelines for planners and decision makers.* ESCAP, Environment and Development Series. UN-ESCAP, Bangkok, 198p.

Estudio R.A. & Gonzales C.L. (1984) Red tides and paralytic shellfish poisoning in the Philippines. In *Toxic Red Tides and Shellfish Toxicity in Southeast Asia* (Ed. by A.W. White, M. Anraku & K.K. Hooi). SEAFDEC and IDRC, Singapore.

Everett G.V. (1973) The rainbow trout *Salmo gairdneri* (Rich.) fishery in Lake Titicaca. *J. Fish. Biol.*, **5**, 429–40.

Falsted K.T., Gjedren T. & Gjerde B. (1993) Genetic improvement of disease resistance in fish: an overview. *Aquaculture* **111**, 65–74.

FAO (1977*a*) Control of the spread of major communicable fish diseases. Report of the FAO/OIE Government Consultation on an international convention for the control of the spread of major communicable fish diseases. *FAO Fish. Rep.*, **192**, 44p.

FAO (1977*b*) China: Recycling of organic wastes in agriculture. *FAO Soils Bull.*, **40**, 107p.

FAO (1980) Report of the fifth session of the IPFC Working Party on Aquaculture and Environment. Jakarta, Indonesia, 22–26 Jan. 1990. *FAO Fish. Rep.*, **241**, 12p.

FAO (1982) Management and utilization of mangroves in Asia and the Pacific. *FAO Environ. Pap.*, No. 3. FAO, Rome, 160p.

FAO (1990) FAO activities related to environment and sustainable development. FAO Council Document CL98/6, Sept. 1990.

FAO (1991) *The Den Bosch Declaration and Agenda for Action on Sustainable Aquaculture and Rural Development.* Conference report and draft proposal of elements for strategies and agenda for action. FAO, Rome.

FAO (1995) *Code of Conduct for Responsible Fisheries.* FAO, Rome.

FAO (1999) *Marine Ranching: Global Perspectives with Emphasis on the Japanese Experience.* FAO Fisheries Circular No. 943. FAO, Rome.

FAO/NACA (1995) *Regional Study and Workshop on Environmental Assessment and Management of Aquaculture Development.* NACA Environment and Aquaculture Development Series No. 1. FAO, Rome; NACA, Bangkok.

FAO/NACA (2000) *The Asia Regional Technical Guidelines on Health Management for the Responsible Movement of Live Aquatic Animals and the Beijing Consensus and Implementation Strategy.* FAO Fisheries Technical Paper No. 402. FAO, Rome, 53p.

FAO/UNEP (1981) Conservation of the genetic resources of fish: problems and recommendations. Report of the Expert Consultation on the Genetic Resources of Fish, Rome, 9–13 June 1980. *FAO Fish. Tech. Pap.*, **217**, 43p.

Fauré A. (1977) Mise au point sur la pollution engendrée les piscicultures. *Piscic. Fr.*, **50**, 33–5.

Feuillet-Girard *et al.* (1988) Nitrogenous compounds in the water column and at the sediment–water interface in the estuarine bay of Mareness-Oleron: Influence of oyster farming. *Aquat. Living Resour.* 1(4), 251–65.

FFI (2002) Low-cost turbot recirculation. *Fish Farming International* **29** (February).

Figueras A.J. (1989) Mussel culture in Spain and France. *World Aquaculture*, **20**(4), 8–17.

Fish G.R. (1966) An artificially maintained trout population in the Northland Lake. *New Zealand J. Sci.*, **9**, 200–10.

Fitzgerald W.J. (1997) Silvofisheries—an environmentally sensitive integrated mangrove forest and aquaculture system. *Aquaculture Asia*, **2**, 9–17.

Floderus S. & Håkanson L. (1987) Resuspension, lenses and nitrogen cycling in Laholmsbukten, S.E. Kattegat. Paper presented at the 4th International Symposium on Sediment/Water Interactions, Melbourne, Feb. 1987.

Folke C. & Kautsky N. (1989) The role of ecosystems for a sustainable development of aquaculture. *Ambio*, **18**(4), 234–43.

Fortes M.D. (1988) Mangrove and seagrass beds of East Asia: habitats under stress. *Ambio*, **17**(3), 207–13.

Fraga S. & Sanchez F.J. (1985) Toxic and potentially toxic dinoflagellates found in Galician Rias (NW Spain). In *Toxic Dinoflagellates* (Ed. by A.W. White & D.G. Baden). Elsevier, New York, 51–4.

Freeman K. (1988) Ecology and aquaculture: shall the twain meet? *Bull. Aquaculture Assoc. Canada*, **88**(2), 82–7.

Friberg L. *et al.* (Eds.) (1974) *Cadmium in the environment.* (Second edition), CRC Press, Cleveland, Ohio.

Fuerst M. (1977) Introduction of *Pacifastacus lenlusculus* (Dana) in Sweden. In *Freshwater Crayfish*. Third International Crayfish Symposium on Freshwater Crayfish (Ed. by O.V. Lindquist). University of Kuopio, Finland, 60–78.

Furfari S.A. (1979) Shellfish purification. In *Advances in Aquaculture* (Ed. by T.V.R. Pillay & W.A. Dill). Fishing News Books, Oxford, England. 385–94.

Gedney R.H., Shang Y.C. & Cook H.L. (1984) Comparative study of tidal and pumped water supply for brackishwater aquaculture ponds in Malaysia. Field Document, Coastal Aquaculture Demonstration and Training Project FI:DP/ MAL/77/008, No. 2, 119–60.

Geldrich E.E. & Clarke N.A. (1966) Bacterial pollution indicators in the intestinal tract of freshwater fish. *Appl. Microbial.*, **14**, 429–37.

GESAMP (1986) *Environmental Capacity: an Approach to Marine Pollution Prevention.* Report of Study Group, GESAMP 30. IMO, London.

GESAMP (1991) *Reducing Environmental Impacts of Coastal Aquaculture.* Report of Study Group, GESAMP 47. IMO, London.

GESAMP (1996*a*) *Monitoring the Ecological Effects of Coastal Aquacultural Wastes.* Report of Study Group GESAMP 57. IMO, London.

GESAMP (1996*b*) *The Contributions of Science to Integrated Coastal Management.* Report of Study Group, GESAMP 61. IMO, London.

GESAMP (2001) *Planning and Management for Sustainable Coastal Aquaculture Development.* Report of Study Group, GESAMP 68. IMO, London.

Getchell R. (1988) Environmental effects of salmon farming. *Aquacult. Mag.*, **14**(6), 44–7.

Ghosh A., Rao L.H. & Banerjee S.C. (1974) Studies on hydrobiological conditions of a sewage-fed pond, with a note on their role in fish culture. *J. Inland Fish. Soc. India*, **6**, 51–61.

Gjedren T. (1985) Improvement of productivity through breeding schemes. *Geological Journal* **10**, 233–41.

Goforth H.W. & Thomas J.R. (1980) Plantings of red mangroves (*Rhizophora mangle* L.) for stabilization of marl shorelines in the Florida Keys. In *Proceedings of the Sixth Annual Conference on the Restoration and Creation of Wetlands* (Ed. by D.P. Cole). Hillsborough Community College, Tampa, Florida, 207.

Gong W.E., Ong J-E. & Wong C.H. (1985) The different uses of mangroves and their possible impacts on mangrove and adjacent coastal fisheries. In *Proceedings of the MAB/COMAR Regional Seminar 'Man's Impact on Coastal and Estuarine Ecosystems'*, Nov. 1984, Tokyo, (Ed. by T. Saski *et al.*). MAB Coordinating Committee of Japan, 21–6.

Gosselink J.G., Odum E.P. & Pope R.M. (1974) *The value of the tidal marsh.* Center for Wetland Resources, Louisiana State University, Baton Rouge, Louisiana, 30p.

Gowen R.J. & Bradbury N.B. (1987) The ecological impact of salmonid farming in coastal waters: A review. In *Oceanography and Marine Biology: An Annual Review* (Ed. by M. Barnes). *Oceanogr. Mar. Biol. Ann. Rev.*, **25**, 563–75.

Grant J. (1996) The relationship of bioenergetics and the environment to the field growth of cultured bivalves. *Journal of Experimental Marine Biology and Ecology* **200**, 239–56.

Grant J. (1999) Ecological constraints on the sustainability of bivalve aquaculture. In *Sustainable Aquaculture—Food for the Future?* (Ed. by N. Svennevig, H. Reinertsen & M. New) 85–96. A.A. Balkema, Rotterdam.

Grant J. & Bacher C. (1998) Comparative models of mussel bioenergetics and their validation at field culture sites. *Journal of Experimental Marine Biology and Ecology* **219**, 21–44.

Gray J. (1990) Antibiotic use in Norwegian fish farms. *Mar. Poll. Bull.*, **21**, 4.

Gregory R.S. (1988) A framework for managing the risks of deliberate releases of genetic material into aquatic ecosystems. *J. Shellfish Res.*, **7**(3), 557.

Grimaldi E. *et al.* (1973) Diffusa infezione branchiale da funghii attribuite al genere *Branchiomyces* Plehn (*Phycomycetes saprolegniales*) a carico dell'ittiofauna di laghi situati a norde a sud delle Alpi. *Mem. Inst. Ital. Idrobiol.*, **30**, 61–96 (in Italian).

Grovhoug J.G. *et al.* (1986) Baseline measurements of butyllin in U.S. harbors and estuaries. In *Oceans '86 Conference Record*, Washington, DC, Sept. 23–5, 1986, 1283–8.

Guélin A. (1962) Polluted water and the contamination of fish. In *Fish as Food, Vol. 2, Nutrition, Sanitation and Utilization* (Ed. by G. Borgstrom). Academic Press, New York, 480–502.

Guerrero R.D. (1982) Ecological impact of fishpens and administrative problems of the fish pen industry. In *Report of the Training Course on Small-scale Pen and Cage Culture for Finfish, Los Banos, Philippines and Aberdeen, Hong Kong, 1981* (Ed. by R.D. Guerrero & V. Soesanto). FAO/UNDP South China Sea Fisheries Development and Coordination Programme, Manila (SCS/PCC/WP-8).

Haavisto P. (1974) *The loading caused by fish farms.* Research Report, National Board of Waters, Helsinki, 74, 79p. (in Finnish).

Håkanson L. *et al.* (1988) *Basic concepts concerning assessments of environmental effects of marine fish farms.* Nordic Council of Ministers, Copenhagen, 103p.

Hall C.B. (1949) *Ponds and fish culture.* Faber & Faber, London, 224p.

Hallerman. E.M. & Kapuscinski A.R. (1990) Transgenic fish and public policy: regulatory concerns. *Fisheries*, **15**(1), 12–20.

Hamilton L.S. & Snedaker S.C. (Eds.) (1984) *Handbook for mangrove area management.* UNEP and Environment and Policy Institute, East West Center, Honolulu, Hawaii, 123p.

Hanamura N. (1979) Advances and problems in culture-based fisheries in Japan. In *Advances in Aquaculture* (Ed. by T.V.R. Pillay & W.A. Dill). Fishing News Books, Oxford, England, 541–7.

Hann J. (1976) Aspects of red mangrove restoration in Florida. In *Proceedings of the Second Annual Conference on the Restoration of Coastal Vegetation in Florida* (Ed. by R.R. Lewis & D.P. Cole). Hillsborough Community College, Tampa, Florida.

Haque M.Z. & Barua G. (1988) Toxic and sub-lethal effect of the Dipteren on *Oreochromis niloticus. Bangladesh J. Fish.*, **11**, 75–9.

Hartman O. *et al.* (1982) Water quality protection on farms with cage rainbow trout culture (English abstract). *Zwocisma Vyroba*, 851–6.

Heald E.J. (1971) *The production of detritus in a South Florida estuary.* Sea Grant Technical Bulletin (University of Miami), No. 6, 110p.

Heald E.J. & Odum W.E. (1970) The contribution of mangrove swamps to Florida fisheries. *Proc. Gulf & Caribb. Fish. Inst.*, 22nd Annu. Sess., 130–5.

Hemmert W.H. (1975) The public health implications of *Gymnodinium breve* red tides, a review of the literature and recent events. In *Proceedings of the First International Conference on Toxic Dinoflagellate Blooms* (Ed. by V.R. Lo Cicero). Massachusetts Science and Technology Foundation, Wakefield, Massachusetts, USA, 489–97.

Henderson J.P. & Bromage N. (1987a) A cost effective strategy for effluent. *Fish Farmer*, **10**(3), 42.

Henderson J.P. & Bromage N. (1987b) Diets to curb pollution. *Fish Farmer*, **10**(4), 39.

Henderson J.P. & Bromage N. (1988) Optimising the removal of suspended solids from aquacultural effluents in settlement lakes. *Aquacult. Eng.*, **7**(5), 167–81.

Hepher B. (1958) On the dynamics of phosphorus added to fish ponds in Israel. *Limn. Oceanogr.* **3**(1), 84–100.

Hepher B. & Schroeder G.L. (1977) Wastewater utilization in integrated aquaculture. In *Wastewater Renovation and Reuse* (Ed. by D'Itri). Marcel Dekker, New York, 529–59.

Hepher B. & Pruginin Y. (1981) *Commercial fish farming.* John Wiley and Sons, New York, 261p.

Hershberger W.K. *et al.* (1990) Genetic changes in growth of coho salmon (*Oncorhyncus kisutch*) in marine net pens, produced by ten years of selection. *Aquaculture* **85**, 187–97.

Hickling C.F. (1962) *Fish Culture.* Faber and Faber, London, 295p.

Hinshaw R.N. (1973) *Pollution as a result of fish cultural activities.* Ecological Research Series, Environmental Protection Agency (USA), Washington, DC 20460, 53p.

Hoffman W.E. & Rogers A.J. (1981) Cost–benefit aspects of coastal vegetation establishment in Tampa Bay, Florida. *Environ. Conserv.*, **8**(1), 39p.

Hopkins J.S. (1996) Aquaculture sustainability: avoiding the pitfalls of the green revolution. *World Aquaculture*, **27**(2), 13–15.

Horna Zapata R.R. (1980) *Reaccion, suelo y mangle.* Unesco Seminar, Cali, Columbia, Nov. 1980.

Horna Zapata R.R. (1981) Reforestacion del mangle (*Rhizophora*). ESPOL Seminar, Guayaquil, Ecuador, 18–27 May 1981.

Howgate P. & Hume A. (1986) Product quality criteria and quality control. In *Realism in Aquaculture: Achievements, Constraints, Perspectives* (Ed. by M. Bilio, H. Rosenthal & C.J. Sindermann). European Aquaculture Society, Bredene, Belgium, 565–76.

Howarth W. (1990) *The Law of Aquaculture.* Fishing News Books, Oxford, England, 304p.

Howarth W. (1999) Legislation for sustainable aquaculture: A legal perspective on the improvement of the Holmenkollen guidelines. In *Sustainable Aquaculture —Food for the Future?* (Ed. by N. Svennevig, H. Reinertsen & M. New) 319–34. A.A. Balkema, Rotterdam.

Huet M. (1986) *Textbook of Fish Culture, 2nd edition.* Fishing News Books, Oxford, England, 438p.

Hungspreugs M. & Yuangthong C. (1984) The present levels of heavy metals in some molluscs of the Upper Gulf of Thailand. *Water Air Soil Pollut.*, **22**, 395–402.

Hungspreugs M. *et al.* (1989) The present status of the aquatic environment of Thailand. *Mar. Pollut. Bull.*, **20**, 327–32.

Hunt J.W. *et al.* (1995) Interactive effects of polyculture, feeding rate and stocking density on growth of juvenile shellfish. *Journal of Shellfish Research* **14**, 191–7.

Huschenbeth E. & Harms U. (1975) On the accumulation of organochlorine pesticides, PCB and certain heavy metals in fish and shellfish from Thai coastal and inland waters. *Arch. Fisch Wiss.*, **26**, 109–22.

ICES (1972) Report of the Working Group on Introductions of Non-indigenous Marine Organisms. *ICES Coop. Res. Rep.*, **32**, 39p.

ICES (1979) Report of the ICES Working Group on the Introduction of Non-indigenous Marine Organisms, ICES/CM: 1979/E22, 37p.

ICES (1982) Status (1980) of introductions of non-indigenous marine species to North Atlantic waters (amendments and additions to Cooperative Research Report No. 32, 1972). *ICES Coop. Res. Rep.*, 86p.

ICES (1984) Guidelines for implementing the ICES code of practice concerning introductions and transfers of marine species. *ICES Coop. Res. Rep.*, **130**, 1–20.

INFOFISH (1997) Sustainable Aquaculture. In *Aquaculture '96* (Ed. by K.P.P. Nambiar & T. Singh). Infofish, Kuala Lumpur.

Ingram M. (1988) Farming of rainbow trout in freshwater tanks and ponds. In *Salmon and Trout Farming* (Ed. by L.M. Laird & T. Needham). Ellis Horwood, Chichester, 155–89.

Inone H. (1972) On water exchange in a net cage stocked with the fish himachi. *Bull. Jpn. Soc. Fish.*, **38**, 167–76.

Isaksson A. (1988) Salmon ranching: a world review. *Aquaculture* **75**, 1–33.

Ito S. & Imai T. (1955) Ecology of oyster bed on the decline productivity due to repeated cultures. *Tokyo J. Agric. Res.*, **4**, 251–68.

Ivasik V.M., Kulakovskaya O.P. & Vorona N.I. (1969) Parasite exchange of herbivorous fish species and carps in ponds of the western Ukraine. *Hydrobiol. J.* **5**(5), 68–71.

Ives B.H. (1989) Pollution! What's a farmer to do? *World Aquaculture*, **20**(3), 48–9.

Jackson P.D. (1981) Trout introduced into south-eastern Australia. Their interaction with native fishes. *Victoria Naturalist*, **98**, 18–24.

Jamieson G.S. & Chandler R.A. (1983) Paralytic shellfish poison in sea scallops (*Placopecten magellanicus*) in West Atlantic. *Canadian J Fish & Aquat. Sci.*, **40**, 313–8.

Janssen W.A. (1970) Fish as potential vectors of human bacterial diseases. *Spec. Publ. Amer. Fish. Soc.*, **5**, 284–90.

Jara R.S. (1984) Aquaculture and mangroves in the Philippines. In *Proceedings of the Workshop on Productivity of the Mangrove Ecosystem: Management Implications* (Ed. by J.E. Ong & W.-K. Gong). Universiti Sains Malaysia, Penang, 97–107.

Jara R.S. (1987) Utilization and management of mangrove resources. In *Papers contributed to the Workshop on Strategies for the Management of Fisheries and Aquaculture in Mangrove Ecosystems, Bangkok, Thailand, June 1986* (Ed. by R.H. Mepham). *FAO Fish. Rep.* (**370**) Suppl., 105–24.

Jefferies D.J., Green J. & Green R. (1984) *Commercial fish and crustacean traps: a serious cause of otter Lutra lutra* (L) *mortality in Britain and Europe*. The Vincent Wildlife Trust, London, 31p.

Jhingran V.G. & Natarajan A.V. (1979) Improvement of fishery resources in inland waters through stocking. In *Advances in Aquaculture* (Ed. by T.V.R. Pillay & W.A. Dill). Fishing News Books, Oxford, England, 532–41.

Johnstone R., Mcintosh D.J. & Wright R.S. (1983) Elimination of orally-administered 17 (alpha)-methyltestosterone by *Oreochromis mossambicus* (Tilapia) and *Salmo gairdneri* (Rainbow trout) juveniles. *Aquaculture*, **35**, 249–57.

Jones K.J. *et al.* (1982) A red tide of *Gyrodinium aureolum* in sea lochs of the Firth of Clyde and associated mortality of pond-reared salmon. *J. Mar. Biol. Assoc. UK*, **62**, 771–82.

Jørgensen O.H. (1982) Legislative measures concerning trout farming in Denmark. In *Report of the EIFAC Workshop on Fish-farm Effluents, 1981* (Ed. by J.S. Alabaster). *EIFAC Tech. Pap.*, **41**, 137–39.

Kalbe L. (1984) Intensive carp rearing in lakes and eutrophication. *Acta Hydrochim. Hydrobiol*, **12**(2), 145–52.

Kannan N. *et al.* (1989) Persistency of highly toxic coplanar PCBs in aquatic ecosystems: uptake and release kinetics of coplanar PCBs in green-lipped mussels (*Perna viridis* Linnaeus). *Environ. Pollut.*, **56**, 65–76.

Kapetsky J.M. (1982) *Some potential environmental effects of coastal aquaculture with implications for site selection and aquaculture engineering*. FAO/UNDP South China Sea Fishery Development and Coordination Programme, Manila. SCS/GEN/82/42, 76–82.

Kapetsky J.M. (1985) Mangrove, fisheries and aquaculture. *FAO Fish. Rep.* (**338**) Suppl. 17–36.

Kapetsky, J.M. (1987) Conversion of mangroves for pond aquaculture: Some short-term and long-term remedies. In *Papers contributed to the Workshop on Strategies for the Management of Fisheries and Aquaculture in Mangrove Ecosystems, Bangkok, Thailand, June 1986* (Ed. by R.H. Mepham). *FAO Fish. Rep.* (**370**) Suppl., 129–41.

Kapuscinski A.R. & Hallerman E.M. (1990) Transgenic fish and public policy. *Fisheries*, **15**(1), 2–11.

Kaspar H.F. *et al.* (1985) Effects of mussel aquaculture on the nitrogen cycle and benthic communities in Kenepuru Sound, Marlborough Sound, New Zealand. *Mar. Biol.*, **85**, 127–56.

Kaushik S.J. *et al.* (1995) Partial or total replacement of fish meal by soya-bean protein, on growth, protein utilisation, potential estrogenic or antigenic effects, cholestromia, and flesh quality, in rainbow trout, *Oncorhyncus mykiss. Aquaculture* **133**, 254–74.

Ketchum B.H. (Ed.) (1972) *The water's edge: Critical problems of the coastal zone.* MIT Press, Cambridge, Massachusetts, 393p.

Ketola G.H. (1982) Effects of phosphorus in trout diets on water pollution. *Salmonid*, July–August 1982, 12–5.

Keup L.E. (1981) Wastewater aquaculture in the United States: Potentials and constraints. In *Aquaculture in Heated Effluents and Recirculation Systems, Vol. 1* (Ed. by C. Tiews). Schriften der Bundesforschungsanstalt für Fischerei, Hamburg, 481–91.

Kilambi R.V. (1981) Cage culture fish production and effects on resident large-mouth bass. In *Aquaculture in Heated Effluents and Recirculation Systems, Vol. 1* (Ed. by K. Tiews). Schriften der Bundesforschungsanstalt für Fischerei, Hamburg, 191–203.

Kilambi R.V. *et al.* (1976) Effects of cage culture fish production upon the biotic and abiotic environment of Crystal Lake, Arkansas. In *Final Report, NOAA NMFS PL88–309 Project*, Department of Zoology, University of Arkansas, 127p.

Kírchner W.B. & Dillon P.J. (1975) An empirical method of estimating retention of phosphorus in lakes. *Water Resources Res.*, **11**, 182–3.

Klemanowicz K.J. & Steele G.H. (1984) Effects of a mechanical oyster harvester on macrobenthic community. *J. Shellfish Res.*, **4**(1), 92.

Klontz G.W., Brock I.R. & MacNair J.A. (1978) *Aquaculture Techniques—Water Use and Discharge Quality.* University of Idaho, Office of Water Research and Technology, Moscow, Idaho, Project A.054-1 DA, 88p.

Kohler C.C. & Stanley J.G. (1984) A suggested protocol for evaluating proposed exotic fish introductions in the United States. In *Distribution, Biology and Management of Exotic Fishes* (Ed. by W.R. Courtenay, Jr. & J.R. Stauffer, Jr.). Johns Hopkins University Press, Baltimore, 387–406.

Korycka A. & Zdanowski B. (1981) Some aspects of the effect of cage fish culture on lakes with special reference to heated lakes. In *Aquaculture in Heated Effluents and Recirculation Systems, Vol. 1* (Ed. by K. Tiews). Schriften der Bundesforschungsanstalt für Fischerei, Hamburg, 131–8.

Korzeniewski K. & Korzeniewska J. (1982) Changes in the composition and physiological properties of the bacterial flora of water and bottom sediments in Lake Letowo caused by intensive trout culture. *Polskie Arch. Hydrobiol.*, **29**(3–4), 671–82.

Krom M.D. & Neori A. (1989) A total nutrient budget for an experimental intensive fish pond with circularly moving seawater. *Aquaculture*, **83**, 345–58.

Krumholz L.A. (1948) Reproduction in the western mosquito fish *Gambusia affinis* (Baerd and Giraud) and its use in mosquito control. *Ecological Monographs*, **18**, 1–43.

Kumagai M. *et al.* (1986) Okadaic acid as the causative toxin of diarrhetic shellfish poisoning in Europe. *Agric. Biol. Chem.*, **50**, 2853–7.

Lahman E.J., Snedaker S.C. & Brown M.S. (1987) Structural comparisons of mangrove forests near shrimp ponds in southern Ecuador. *Interciencia*, **12**(5), 240–3.

Landesman L. (1994) Negative impacts of coastal aquaculture development. *World Aquaculture* **25**, 12–5.

Larsson A.M. (1984) *Hydrological and chemical observations in a coastal area with mussel farming, W. Sweden.* University of Gothenberg, Dept. of Oceanography, Report 46, 29p.

Larsson A.M. (1985) Blue mussel sea farming—effects on water quality. *Vatten*, **41**(4), 218–24.

Lasordo T.M., Masser M.P. & Rakocy J. (2001) Recirculating aquaculture tank production systems. *World Aquaculture* **32**, 18–22.

Lauenstein G.G. & O'Connor T.P. (1989) Organic chemical contamination in mollusks of the coastal and estuarine United States. *World Aquaculture*, **20**(4), 101–3.

Lee C.S., Sweeney J.N. & Richards Jr. W.K. (1986) Marine shrimp aquaculture: a novel waste treatment system. *Aquacult. Eng.*, **5**, 147–60.

Lee J.S. *et al.* (1988) Diarrhetic shellfish toxins in Norwegian mussels. *Nippon Suisan Gakkaishi*, **54**, 1953–7.

Lee J.S. *et al.* (1989) Determination of diarrhetic shellfish toxins in various dinoflagellate species. *J. Appl. Phycol.*, **1**, 147–52.

Leffertstra H. (1988) *Fish farming and eutrophication of the North Sea.* Norwegian State Pollution Control Authority.

Leffertstra H. & Kryvi H. (1988) Report on suspended solids from fish farms in Norway. In *National contributions on suspended solids from land-based fish farms* (Ed. by M. Pursiainen). EIFAC Working Party on Fish Farm Effluents, 1st Session, The Hague.

Leopold L. (1969) Method for measuring landscape appeal. *Natural History*, **8**, 36–45.

Leopold L.B. *et al.* (1971) *A procedure for evaluating environmental impacts.* U.G. Geological Survey Circ. No. 645. Govt. Printing Office, Washington, DC.

Lewis R.R. (1982) Mangrove forests. In *Creation and Restoration of Coastal Plant Communities* (Ed. by R.R. Lewis). CRC Press, Boca Raton, Florida, 153–71.

Liao P. (1970a) Pollutional potential of salmonid fish hatcheries. *Water Sewage Works*, **117**(8), 291–7.

Liao P. (1970b) Salmonid hatchery wastewater treatment. *Water Sewage Works*, **117**(12), 439–43.

Liao P.B. (1971) Water requirements of salmonids. *Prog. Fish. Cult.* **33**, 210–15.

Liao P.B. & Mayo R.D. (1974) Intensified fish culture combining water reconditioning with pollution abatement. *Aquaculture*, **3**, 61–85.

Lovell R.T. (1979) Flavour problems in fish culture. In *Advances in Aquaculture* (Ed. by T.V.R. Pillay & W.A. Dill). Fishing News Books, Oxford, England, 186–90.

Lovshin L.L. *et al.* (1986) *Cooperatively managed rural Panamanian fish ponds: the integrated approach.* Research and Development Series No. 2, International Center for Aquaculture, Auburn University, Auburn, Alabama.

Lugo A.E. and Cintron G. (1975) The mangrove forests of Puerto Rico and their management. In *Proceedings of the International Symposium on the Biology and Management of Mangroves* (Ed. by G.E. Walsh, S. Snedakar & H. Teas). Institute of Food and Agriculture Sciences, University of Florida, Gainsville, Florida, 825–46.

Lugo A.E. & Snedaker S.C. (1974) The ecology of mangroves. *Annual Review of Ecology and Systematics*, **5**, 39–64.

Macintosh D.J. (1982) Fisheries and aquaculture significance of mangrove swamps, with special reference to the Indo-West Pacific region. In *Recent*

Advances in Aquaculture, Vol. 1 (Ed. by J.F. Muir & R.J. Roberts). Croom Helm, London, 4–85.

Macnae W. (1967) Zonation within mangroves associated with estuaries in North Queensland. In *Estuaries* (Ed. by G.H. Lauff). Publication No. 83, American Association for the Advancement of Science, Washington D.C., 432–41.

Macnae W. (1968) A general account of the fauna and flora of mangrove swamps and forests in the Indo-West Pacific region. *Adv. Mar. Biol.*, **6**, 73–270.

Macnae, W. (1974) *Mangrove forests and fisheries.* Indian Ocean Programme Publication No. 34. IOFC, Rome. IOFC/Dev/74/34, 35p.

Maguire R.J. (1986) Review of the occurrence, persistence and degradation of the tributyltin in freshwater ecosystems in Canada. In *Oceans '86 Conference Record*, Washington, DC, Sept. 23–5, 1986, 1252–5.

Maine P.D. & Nash C.E. (1987) *Aquaculture sector development—A guideline for the preparation of a national plan.* ADCP/REP/87/27, UNDP/FAO, Rome, 21p.

Mäkinen T., Lindgren S. & Eskelinen P. (1988) Sieving as an effluent treatment for aquaculture. *Aquacult. Eng.*, **7**, 367–77.

Mangelson K.A. & Watters G.Z. (1972) Treatment efficiency of waste stabilisation ponds. *J. San. Eng. Div., Proc. A.S.C.E.* (April), 407–25.

Mann R. (Ed.) (1979) *Exotic species in aquaculture.* MIT Press, Cambridge, USA, 106–22.

Mantle G. (1980) Coping with pollution. *Fish Farmer*, **2**, 38–41.

Mantle G.J. (1982) Biological and chemical changes associated with the discharge of fish farm effluent. In *Report of the EIFAC Workshop on Fish Farm Effluents* (Ed. by J.S. Alabaster). *EIFAC Tech. Pap.*, **41**, 103–12.

Markmann P.N. (1978) Begroensning af dambrugsforureningen. *Vand*, **9**(1), 29–34.

Markmann P.N. (1982) Biological effects of effluents from Danish fish farms. In *Report of the EIFAC Workshop on Fish Farm Effluents* (Ed. by J.S. Alabaster). *EIFAC Tech. Pap.*, **41**, 99–102.

Martin J. (1987) Sites for sore eyes. *Fish Farmer*, **10**(4), 17–18.

Martosubroto P. & Naamin H. (1977) Relationship between tidal forests (mangroves) and commercial shrimp production in Indonesia. *Mar. Res. Indonesia*, **18**, 81–6.

Mattsson J. & Linden O. (1983) Benthic microfauna succession under mussels, *Mytilus edulis*, cultured on hanging long lines. *Sarsia*, **68**, 97–102.

Mayo R.D. (1981) Recirculation systems in North America. In *Aquaculture in Heated Water Effluents and Recirculation Systems, Vol. II* (Ed. by K. Tiews). Schriften der Bundesforschungsanstalt für Fischerei, Hamburg, 329–42.

McAtee W.L. & Piper S.E. (1937) Excluding birds from reservoirs and fish ponds. US Dept. of Agriculture Leaflet, 120p.

McDowall R.M. (1968) Interaction of the native and alien faunas of New Zealand and the problem of fish introductions. *Trans. Amer. Fish. Soc.*, **97**, 1–11.

McNeil W.J. (1979) Review of transplantations and artificial recruitment of anadromous species. In *Advances in Aquaculture* (Ed. by T.V.R. Pillay & W.A. Dill). Fishing News Books, Oxford, England, 547–54.

Meyer J. (1981) Easy pickings. *Birds Summer*, **8**(6), 51–3.

Meyers T.R. (1980) Experimental pathogenicity of reovirus $13p_2$ for juvenile American oysters *Crassostrea virginica* (Gmelin) and blue-gill fingerlings *Lepomis macrochirus* (Rafinesque). *J. Fish. Dis.*, **6**, 277–92.

Meyers T.R. (1984) Marine bivalve mollusks as reservoirs of viral pathogens: significance to marine and anadromous finfish aquaculture. *Mar. Fish. Rev.*, **46**(3), 14–7.

Michel P. (1980) Methodological problems of impact studies in coastal wetlands: Management of the Mauguio Pond and enlargement of the Rhone–Sète canal. In *First National Meeting. Littoral Wetlands, Aquaculture and Wild Fauna, Montpellier, 17–19 June 1980. Bull. Mens. Off. Natl. Chasse*, 269–75 (in French).

Midtyling P.V. (1985) Use of drugs and disinfectants in Norwegian fish farms. *Vann*, **3**, 177–80 (in Swedish).

Mills S. (1982) Britain's native trout is floundering. *New Scientist*, 25th November 1982, 498–501.

Möller B. (1987) Udledning av organisk stof, kvaelstof og fosfor fra dansk havsbrug 1985. *Dansk Akvakultur Inst.*, **18**.

Molnar J.J., Schwartz N.B. & Lovshin L.L. (1985) Integrated aquaculture development: sociological issues in the cooperative management of community fish ponds. *Sociologia Ruralis*, **15**(1), 61–80.

Moriarty D.J. (1986) Bacterial productivity in ponds used for culture of peneid prawns. *Microbial Ecology*, **12**, 259–70.

Moriarty D.J.W. (1997) Probiotics and biotechnology for sustainable aquaculture. In *Aquaculture '96* (Ed. by K.P.P. Nambiar & T. Singh) 115–21. Infofish, Kuala Lumpur.

Morrison M.L., Slack R.D. & Shanley E. (Jr.) (1978) Age and foraging efficiency of olivaceous cormorants. *Wilson Bulletin*, **90**, 414–22.

Mortensen B.F. (1977) Er rensning af aflob fra dambrug realistisk. *Ferskvandsfiskeribladet*, **75**(1), 2–6.

Mott D.F. (1978) Control of wading bird predation at fish rearing facilities. In *Wading Birds* (Ed. by A. Sprunt IV, J.C. Ogden & S. Winckler). Research Report No. 7. National Audubon Society, New York.

Muir J.F. (1982) Economic aspects of waste treatment in fish culture. In *Report of the EIFAC Workshop on Fish-Farm Effluents, Denmark, May 1981* (Ed. by J.S. Alabaster). *EIFAC Tech. Pap.*, **41**, 123–35.

Mukherjee A.K. and Tiwari K.K. (1984) Mangrove ecosystems under induced stress: A case history of Sunderbans, West Bengal, India. In *Proceedings of the Asian Symposium on Mangrove Environment: Research and Management, August 1980, Kuala Lumpur* (Ed. by E. Soepadmo, A.N. Rao & D.J. Macintosh), 633–43.

Munn R.E. (Ed.) (1979) *Environmental impact assessment—Principles and procedures, SCOPE 5, 2nd Edition*. John Wiley & Sons, Chichester, England.

Munro A.L.S. (1986) Transfers and introductions: do the dangers justify greater public control? In *Realism in Aquaculture: Achievements, Constraints, Perspectives* (Ed. by M. Bilio, H. Rosenthal & C.J. Sindermann). European Aquaculture Society, Bredene, Belgium, 375–94.

Munro A.L.S., Leversidge J. & Elson K.G.R. (1976) The distribution and prevalence of infectious pancreatic necrosis virus in wild fish in Loch Awe. *Proc. Roy. Soc. Edinb.* (B). **75**, 223–32.

Munro A.L.S. *et al.* (1980) The quarantine of coho salmon for a whole life cycle. *ICES 1980/E*, **51**, 1–7.

Murphy J.P. & Lipper R.I. (1970) BOD production of channel catfish. *Prog. Fish-Cult.*, **32**(4), 195–8.

Nash C.E. & Brown C.M. (1980) A theoretical comparison of waste treatment processing ponds and fish production ponds receiving animal wastes. In *Integrated Agriculture–Aquaculture Farming Systems* (Ed. by R.S.V. Pullin & Z.H. Shehadeh). *ICLARM Conf. Proc.*, **4**, 87–97.

Nash C.E. & Paulsen C.L. (1981) Water quality changes relevant to heated effluents and intensive aquaculture. In *Aquaculture in Heated Effluents and Recirculation Systems, Vol. I* (Ed. by C. Tiews). Schriften der bundesforschungsanstalt für Fischerei, Hamburg, 3–15.

NATS (1998) *Holmenkollen Guidelines for Sustainable Aquaculture*. Second International Symposium on Sustainable Aquaculture, Oslo. Norwegian Academy of Technological Sciences, Trondheim, Norway.

Nauen C.C. (1983) *Compilation of legal limits for hazardous substances in fish and fishery products*. FAO of the UN, Rome, Italy.

NCC (1989) *Fish Farming and the Safeguard of the Marine Environment of Scotland*. Based on a report prepared by the Institute of Aquaculture, University of Stirling. Nature Conservancy Council, Peterborough, 136p.

Needham T. (1984) Are farmed salmon a genetic threat to wild stocks? *Fish Farmer*, **7**(6), 35.

Negroni G. (2000) Management optimization and sustainable technologies for treatment and disposal/reuse of fish farm effluent with emphasis on constructed wetlands. *World Aquaculture* **31**, 16–63.

Neori A. *et al.* (1996) Seaweed biofilters as regulators of water quality in integrated fish sea weed culture units. *Aquaculture* **141**, 183–99.

NERC (1975) *Flood studies report*. Natural Environment Research Council, London.

Newkirk G.F. (1980) Genetic aspects of the introduction and culture of non-indigenous species for aquaculture. In *Exotic Species in Mariculture* (Ed. by R. Mann). MIT Press, Cambridge, USA, 192–211.

Niemi M. & Taipalinen I. (1980) Hygienic indicator bacteria in fish farms. *Research Report, National Board of Waters, Helsinki*, **14** (in Finnish).

Niemi M. & Taipalinen I. (1982) Faecal indicator bacteria at fish farms. *Hydrobiologia*, **86**, 171–5.

Nishitani L. & Chew K. (1988) PSP toxins in the Pacific coast states: monitoring programs and effects on bivalve industries. *J. Shellfish Res.*, **7**, 653–69.

Nixon S.W. (1980) Between coastal marshes and coastal waters—a review of twenty years of speculation and research on the role of salt marshes in estuarine productivity and water chemistry. In *Estuarine Wetland Processes* (Ed. by P. Hamilton & K.B. Macdonald). Plenum Publishing Corp., New York, 437–520.

Nji A. (1986) *Social, cultural and economic determinants of the adoption of fish pond culture in Menoua, Mezam and Momo divisions of Cameroon*. Dschang University Center, Dschang, Cameroon.

Noakes D.S.P. (1955) Methods of increasing growth and obtaining natural regeneration of the mangrove type in Malaya. *Malay. For.*, **18**, 22.

Nor S.M. (1984) Major threats to the mangroves of Asia and Oceania. In *Proceedings of the Workshop on Productivity of the Mangrove Ecosystem: Management Implications* (Ed. by J.-E. Ong and W.-K. Gong). Universiti Sains Malaysia, Penang, 69–78.

NORAD/UNDP/FAO (1987) *Women in aquaculture—proceedings of the ADCP/NORAD workshop on women in aquaculture, Rome, FAO, 13–16 April 1987* (Ed. by C.E. Nash, C.R. Engle & D. Crosetti). ADCP/REP/87/28, FAO, Rome, 123p.

Odum E.P. (1968) A research challenge: evaluating the productivity of coastal and estuarine water. In *Proceedings of the Second Sea Grant Conference*, University of Rhode Island, Kingston, University of Rhode Island, 63–4.

Odum E.P. (1989) *Ecology and Our Endangered Life Support Systems*. Sinauer Associates, Inc. Publ., Massachusetts, 283p.

Odum E.P. & de la Cruz A.A. (1967) Particulate organic detritus in a Georgia salt marsh–estuarine ecosystem. In *Estuaries* (Ed. by G.H. Lauff). Publication No. 83, American Association for the Advancement of Science, Washington, DC, 383–8.

Odum W.E. (1971) Pathways of energy flow in a South Florida estuary. *University of Miami Sea Grant Technical Bulletin*, **7**, 162p.

OECD (1982) *Eutrophication of waters, assessment and control*. Organisation for Economic Cooperation and Development, Paris, 215p.

Ong J.-E. (1984) Mangrove outwelling? In *Proceedings of the Unesco Workshop on productivity of the mangrove ecosystem: management implications* (Ed. by J.-E. Ong and W.-K. Gong). Universiti Sains Malaysia, Penang, 30–6.

Ong J.-E. (1987) Ecological research needs in relation to the impacts of various uses on the mangrove ecosystems of the Indo-Pacific region. In *Papers contributed to the Workshop on Strategies for the Management of Fisheries and Aquaculture in Mangrove Ecosystems, Bangkok, Thailand, June 1986* (Ed. by R.H. Mepham). *FAO Fish, Rep.*, **370**, Suppl., 83–92.

Oswald W.J. (1973) Complete waste treatment in ponds. In *Water Quality Management and Pollution Control Problems. Progress in Water Technology Vol. 3* (Ed. by S.H. Jenkins). Pergamon, New York, 156–63.

Ottman F. & Sornin J.M. (1985) Observations on sediment accumulation as a result of mollusk culture systems in France. In *Proceedings of the International Symposium on Utilization of Coastal Ecosystems: Planning, Pollution and Productivity, 21–27 Nov. 1982, Rio Grande, Brazil Vol. 1* (Ed. by N.L. Chao & W. Kirby-Smith). 329–37.

Pannier F. (1979) Mangroves impacted by human-induced disturbance: A case study of the Orinoco Delta mangrove ecosystem. *Environmental Management,* **3**, 205–16.

Parks R.W., Scarsbrook E. & Boyd C.E. (1975) Phytoplankton and water quality in a fertilized fish pond. *Auburn Univ. Agric. Exp. Stn. Circ.,* **224**, 16p.

Parson D.A. (1949) The hydrology of a small area near Auburn, Alabama. *U.S. Soil Conserv. Ser.,* **40**.

Patil P.G. & Krishnan M. (1998) The social impacts of shrimp farming in Nellore District, India. *Aquaculture Asia,* **3**(1), 3–5.

Paul J.D. & Davies I.M. (1986) Effects of copper and tin-based antifouling compounds on the growth of scallops (*Pecten maximus*) and oysters (*Crassostrea gigas*), *Aquaculture,* **54**, 191–203.

Penczak T. *et al.* (1982) The enrichment of a mesotrophic lake by carbon, phosphorus and nitrogen from the cage aquaculture of rainbow trout, *Salmo gairdneri. J. Appl. Ecol.* **19**, 371–93.

Penn G.H. (1954) Introductions of American crayfishes into foreign lands. *Ecology,* **35**, 296.

Persson G. (1988) Environmental impact by nutrient emissions from salmonid fish culture. In *Eutrophication and Lake Restoration, Water Quality and Biological Impacts* (Ed. by G. Balvay). Institut Natl. de la Recherche Agronomique, Thonon-les-Bains (France), 215–26.

Peters F. & Neukirch M. (1986) Transmission of some fish pathogenic viruses by the heron *Ardea cinerea. J. Fish Dis.,* **9**, 539–44.

Peterson C.H., Summerson H.C. & Fegley S.R. (1987) Ecological consequences of mechanical harvesting of clams. *Fish. Bull.* **85**(2), 281–98.

Petit J. (1978) Treatment experiments of the water discharged by a fish farm. *Technical Report of the Mission in UK,* Jony en Josas, INRA, 30p.

Phillips D.J.H. (1985) Organochlorines and trace metals in green-lipped mussels *Perna viridis* from Hong Kong waters: a test of indicator ability. *Mar. Ecol. Prog. Ser.,* **21**, 251–8.

Phillips D.J.H. *et al.* (1986) Trace metals in Pacific oysters (*Crassostrea gigas* Thunberg) marketed in Hong Kong. In *The First Asian Fisheries Forum, Vol. 1* (Ed. by J.L. MacLean, L.B. Dixon & L.V. Hosillos). Asian Fisheries Society, Manila.

Phillips M. & Barg V. (1999) Experiences and opportunities in shrimp farming, In *Sustainable Aquaculture—Food for the Future?* (Ed. by N. Svennevig, H. Reinertsen & M. New) 43–72. A.A. Balkema, Rotterdam.

Phillips M. & Beveridge M. (1986) Cages and the effect on water condition. *Fish Farmer,* **9**(3), 17–9.

Phillips M.J., Beveridge M.C.M. & Ross L.G. (1985) The environmental impact of salmonid cage culture in inland fisheries: Present status and future trends. *J. Fish. Biol.,* **27**(A), 123–37.

Phillips M.J., Beveridge M.C.M. & Clarke R.M. (1991) Impact of aquaculture on water resources. In *Aquaculture and Water Quality* (Ed. by D.E. Brune & J.R. Tomasso). *Advances in World Aquaculture* **3**, 568–91.

Phillips M.J., Lin C.K. & Beveridge M.C.M. (1993) Shrimp culture and the environment. In *Environment and Aquaculture in Developing Countries* (Ed. by R.S.V. Pullin, H. Rosental & J.L. MacLean) 171–97. ICLARM Conference Proceedings 31. ICLARM, Manila.

Pillai T.G. (1972) Pests and predators in coastal aquaculture systems of the Indo-Pacific Region. In *Coastal Aquaculture in the Indo-Pacific Region* (Ed. by T.V.R. Pillay). Fishing News Books, Oxford, England, 456–70.

Pillay T.V.R. (1954) The ecology of a brackishwater *bheri* with special reference to the fish cultural practices and biotic interaction. *Proc. Natl. Inst. Sci. India*, **20**(4), 399–427.

Pillay T.V.R. (1958) Land reclamation and fish culture in the deltaic area of West Bengal, India. *Progr. Fish-Cult.*, **20**, 99–103.

Pillay T.V.R. (1965) *Report to the Government of Nigeria on investigations of the possibility of brackishwater fish culture in the Niger Delta*, Rep. FAO/EPTA (1973), 52p.

Pillay T.V.R. (1967) Estuarine fisheries of the Indian Ocean coastal zone. In *Estuaries* (Ed. by G.H. Lauff). Publication No. 83, American Association for the Advancement of Science, Washington, DC, 647–57.

Pillay T.V.R. (1977) *Planning of aquaculture development—an introductory guide.* Fishing News Books, Oxford, England, 71p.

Pillay T.V.R. (1990) *Aquaculture: Principles and Practices.* Fishing News Books, Oxford, UK, 575p.

Pillay T.V.R. (1996) The challenges of sustainable aquaculture. *World Aquaculture*, **27**(2), 7–9.

Pillay T.V.R. (1997) Aquaculture development and the concept of sustainability. In *Proceedings of the INFOFISH—Aquatech '96* (Ed. by K.P.P. Nambiar & T. Singh) 1–6. International Conference on Aquaculture, Kuala Lumpur, Malaysia.

Pillay T.V.R. (1999) Resources and constraints for sustainable aquaculture. In *Sustainable Aquaculture—Food for the Future?* (Ed. by N. Svennevig, H. Reinertsen & M. New) 21–7. A.A. Balkema, Rotterdam.

Pillay T.V.R., Dutta S.N. & Rajagopal S. (1954) The vibrio flora of fishes, water and silt in the Hooghly estuary, with reference to cholera endemicity. *Al. Assoc. Bull.*, All-India Institute of Hygiene and Public Health, Calcutta, Jan. 1954, 1–5.

Piyakarnchana T. (1980) *The present state of mangrove ecosystems in Southeast Asia and the impact of pollution—Thailand.* FAO/UNEP, SCS/80/WP/94 (Rev.), 21.

du Plessis S.S. (1957) Growth and daily food intake of the white-breasted cormorant in captivity. *The Ostrich*, December 1957, 197–201.

Prakash A., Medcof J.C. & Tennant A.D. (1971) *Paralytic shellfish poisoning in eastern Canada.* Fisheries Research Board of Canada, 177, Ottawa, Canada.

Price G.A. & Clements M.S. (1974) Some lessons from model and full-scale tests in rectangular sedimentation tanks. *Water Pollution Control*, **73**(1), 102–13.

Pullin R.S.V. (1989) Third world aquaculture and the environment. *Naga*, **12**(1), 10–3.

Quayle D. (1965) *Animal detoxification.* Proceedings of Joint Sanitation Seminar on North Pacific Clams, September 24–5, 1965.

Querellou J., Fauré A. & Fauré C. (1982) Pollution loads from rainbow trout farms in Brittany, France. In *Report of the EIFAC Workshop on Fish Farm Effluents, May 1981* (Ed. by J.S. Alabaster). *EIFAC Tech. Pap.*, **41**, 87–97.

Quinney T.E. & Smith P.C. (1980) Comparative foraging behaviour and efficiency of adult and juvenile great blue herons. *Can. J. Zool.*, **58**, 1168–73.

Rabinowitz D. (1975) Planning experiments in mangrove swamps of Panama. In *Proceedings of the International Symposium on the Biology and Management of Mangroves, Vol. 1* (Ed. by G.E. Walsh, S.C. Snedaker & H.J. Teas). Institute of Food and Agricultural Sciences, University of Florida, Gainseville, Florida, 385–93.

Rakocy J.E. (1989) Vegetable hydroponics and fish culture—a productive interface. *World Aquaculture*, **20**(3), 42–7.

Rao D.V.S., Quilliam M.A. and Pocklington R. (1988) Demoic acid—A neurotoxic amino acid produced by the marine diatom *Nitzchia pungens* in culture. *Canadian J. Fish & Aquat. Sci.*, **45**, 2076–9.

Rasmussen F. (1988) *Therapeutics used in fish production: pharmokinetics, residues and withdrawal periods.* EIFAC/XV/88/Inf. 13, FAO, Rome, 22p.

Reckhow K.H. & Chapra S.C. (1983) *Engineering approaches for lake management, Vol. 1: Data analysis and empirical modelling.* Ann Arbor Science Publishers, Ann Arbor, MI.

Roberts J.K. (1989) Low pollution feeds. *Trout News*, 8th June, 22–7.

Roberts R.J. & Muir J.F. (1995) 25 years of world aquaculture: Sustainability, a global program. In *Sustainable Fish Farming* (Ed. by H. Reinertsen & H. Haaland) 167–181. A.A. Balkema, Rotterdam/Brookfield.

Rohlich G.A. (1964) Methods for the removal of phosphorus and nitrogen from sewage plant effluents. In *Water Quality Engineering for Practising Engineers*, by W.W. Eckenfelder Jr. Barnes and Noble, New York, 207–23.

Rollet B. (1984) *La ecologia de los manglares con referencia especial a la base biologica para la ordenacion sostenida, forestal y pesca.* Papers presented at the FAO Seminario regional sobre la ordenacion integrada de los manglares, Cuba–Venezuela–Colombia, November 1984, 33p. (Mimeo.)

Rosenthal H. (1978) Bibliography on transplantation of aquatic organisms and its consequences on aquaculture and ecosystems. *European Maricult. Soc., Spec. Publ.*, **3**, 1–146.

Rosenthal H. (1980) Implications of transplantations to aquaculture and ecosystems. *Mar. Fish. Rev.*, **42**(5), 1–4.

Rosenthal H. (1981) Recirculation systems in Western Europe. In *Aquaculture in Heated Effluents and Recirculation Systems, Vol. 2* (Ed. by K. Tiews). Schriften der Bundesforschungsanstalt für Fischerei, Hamburg, 305–15.

Rosenthal H. & Murray K.R. (1986) System design and water quality criteria. In *Realism in Aquaculture: Achievements, Constraints, Perspectives* (Ed. by M. Bilio, H. Rosenthal & C.J. Sindermann). European Aquaculture Society, Bredene, Belgium, 473–93.

Rosenthal H., Weston D., Gowen R. & Black E. (Eds.) (1988) Report of the *ad hoc* Study Group on the 'environmental impact of mariculture'. *ICES. Coop. Res. Rep.*, **154**, 83p.

Ross A. (1988) *Controlling nature's predators on fish farms.* Marine Conservation Society, 96p.

Royal Norwegian Ministry of Fisheries (1985) *Fish Farming—Act of 14 June 1985. No. 68 relating to the breeding of fish, shellfish, etc.* Fiskeridepartmentet Informer, 20p.

Rubino M.C. & Wilson C.A. (1993) Issues in aquaculture regulation, National Oceanic and Atmospheric Administration, October 1993. Bluewater Inc. Bethesda, MD. 29814.

Ruddle K. (1990) *The context of small-scale integrated agriculture-aquaculture systems in Africa: a case study of Malawi.* ICLARM, Manila.

Rychly J. (1980) Nitrogen balance in trout. II. Nitrogen excretion and retention after feeding diets with varying protein and carbohydrate levels. *Aquaculture*, **20**, 343–50.

Rychly J. & Spannhof L. (1979) Nitrogen balance in trout. I. Digestibility of diets containing varying levels of protein and carbohydrate. *Aquaculture*, **16**(1), 39–46.

Saclauso C.A. (1989) Brackishwater aquaculture: Threat to the environment? *Naga*, July, 6–9.

Saenger P., Hegerl E.J. & Davie J.D.S. (Eds.) (1983) Global status of mangrove ecosystems. *IUCN Commission on Ecology Papers*, **3**, 88p.

Samuelson O.B. (1989) Degradation of oxytetracycline in seawater at two different temperatures and light intensities and the persistence of oxytetracycline in the sediment from a fish farm. *Aquaculture*, **83**, 7–16.

Sandifer P.A. (1986) Some recent advances in the culture of crustaceans. In *Realism in Aquaculture: Achievements, Constraints, Perspectives* (Ed. by M. Bilio, H. Rosenthal & C.J. Sindermann). European Aquaculture Society, Bredene, Belgium, 143–71.

Sassaman R.W. (1981) Threshold of concern: A technique for evaluating environmental impacts and amenity values. *J. Forestry*, **79**(2), 84–6.

Schmid A. (1980) Arzneimittelrückstande be fischen. *Tieräztl. prax.*, **8**, 237–44.

Schnick R.A., Meyer F.P. & Walsh D.F. (1986) Status of fishery chemicals in 1985. *Progr. Fish-Cult.* **48**(1), 1–17.

Schotissek C. & Naylor E.R. (1988) Fish farming and influenza pandemics. *Nature*, **331**, 215.

Schroeder G. (1975) Some effects of stocking fish in waste treatment ponds. *Water Res.*, **9**, 591–93.

Schroeder G. & Hepher B. (1979) Use of agricultural and urban wastes in fish culture. In *Advances in Aquaculture* (Ed. by T.V.R. Pillay & W.A. Dill). Fishing News Books, Oxford, England, 487–89.

Schwartz M.R. & Boyd C.E. (1995) The use of constructed wetlands to treat aquacultural effluents. *World Aquaculture* **26**, 42–3.

Scottish Executive (1999) Policy guidance note: locational guidelines for the authorisation of marine fish farms in Scottish waters. *Scottish Executive, Office of the First Minister*, Edinburgh.

Shafer E.L. Jr., Hamilton J.F. Jr. & Schmidt E.A. (1969) Natural landscape preferences: A predictive model. *J. Leisure Res.* **1**(1).

Sharp G.J. & Lamson C. (1988) Approaches to reducing conflict between traditional fisheries and aquaculture. *World Aquaculture*, **20**(1), 79–83.

Shimizu Y. & Yoshioka M. (1981) Transformation of paralytic shellfish toxins as demonstrated in scallop homogenates. *Science*, **212**, 547–9.

Short J.W. & Thrower F.P. (1986a) Tri-n-butyltin caused mortality of chinook salmon, *Oncorhynchus tshawytscha*, on transfer to TBT-treated marine net pen. In *Ocean '86 Conference Record, Organotin Symposium, Vol. 4*, 1202–5.

Short J.W. & Thrower F.P. (1986b) Accumulation of butyltins in muscle tissue of chinook salmon reared in sea pens treated with tri-n-butyltin. In *Oceans '86 Conference Record, Organotin Symposium, Vol. 4*, 1177–81.

Shumway D.L. & Palensky J.R. (1973) Impairment of the flavour of fish by water pollutants. *U.S. Government Report PB-221-480, EPA-R3-73-010*. Environment Protection Agency, Washington, DC.

Shumway S.E., Sherman-Caswell S. & Hurst J.W. (1988) Paralytic shellfish poisoning in Maine: monitoring a monster. *J. Shellfish Res.*, **7**, 643–52.

Shumway S.E. (1990) A review of the effects of algal blooms on shellfish and aquaculture. *J. World Aquacult. Soc.*, **21**(2), 65–104.

Silas E.G. (1987) Management of mangrove associated fisheries and aquaculture in the Sunderbans, India. In *Papers contributed to the Workshop on Strategies for the Management of Fisheries and Aquaculture in Mangrove Ecosystems, Bangkok, Thailand, June, 1986* (Ed. by R.H. Mepham). *FAO Fish. Rep.*, **(370)** Suppl., 21–43.

de Silva S.S. (1999) Feed resources, usage and sustainability In *Sustainable Aquaculture—Food for the Future?* (Ed. by N. Svennevig, H. Reinertsen & M. New) 221–42. A.A. Balkema, Rotterdam.

Sindermann C.J. (1986a) The role of pathology in aquaculture. In *Realism in Aquaculture: Achievements, Constraints, Perspectives* (Ed. by M. Bilio, H. Rosenthal & C.J. Sindermann). European Aquaculture Society, Bredene, Belgium, 395–419.

Sindermann C.J. (1986b) Strategies for reducing risks from introductions of aquatic organisms: A marine perspective. *Fisheries*, **11**(2), 10–15.

Sivalingam P.M. & Bhaskaran B. (1980) Experimental insight of trace metal environmental pollution problems in mussel farming. *Aquaculture*, **20**, 291–303.

Skaala O. *et al.* (1990) Interactions between natural and farmed fish populations: information from genetic markers. *J. Fish. Biol.*, **36**, 449–60.

Sladeck V. (1973) System of water quality from biological point of view. *Arch. Hydrobiol.*, **7**, 218.

Solbe J.F. de L.G. (1982) Fish-farm effluents: A United Kingdom Survey. In *Report of the EIFAC Workshop on Fish Farm Effluents* (Ed. by J.A. Alabaster). *EIFAC Tech. Pap.*, **41**, 29–55.

Solbe J.F. de L.G. (1987) Effluent control and the U.K. fish farmer. English Fish Farming Conference, Hampshire College of Agriculture, 12–13 Sept. 73–85.

Solberg S.O. & Bregnballe F. (1982) Pollution from farmed trout fed with minced trash fish. In *Report of the EIFAC Workshop on Fish Farm Effluents* (Ed. by J.S. Alabaster). *EIFAC Tech. Pap.*, **41**, 65–71.

Sonnenholzner S. & Boyd C.E. (2000) Managing the accumulation of organic matter deposited on the bottom of shrimp ponds—Do chemical and biological probiotics really work? *World Aquaculture* **31**, 24–8.

Sonstegard R.A. (1977) The potential utility of fishes as indicator organisms for environmental carcinogens. In *Wastewater Renovation and Reuse* (Ed. by D'Itri). Marcel Dekker Inc., New York, 561–77.

Sorensen J.C. (1971) *A framework for identification and control of resource degradation and conflict in the multiple use of the coastal zone.* Master's thesis, University of California, Berkeley.

Sornin J.M. (1979) *Enquete sur la sedimentation et l'exhausement des fonds dons les zones conchylicoles des côtes des France.* Rapport DEA, Université de Nantes, 41p.

Sornin J.M. (1981) *Sedimentary processes and biodeposition linked to different shellfish culture methods.* Lab. de Geologie Marine et Appliquee, 188p.

Spanier E. (1979) The use of distress calls to repel night herons (*Nycticorax nycticorax*) from fish ponds. *J. Appl. Ecol.*, **17**, 287–94.

Speece R.E. (1973) Trout metabolism characteristics and the rational design of nitrification facilities for water re-use in hatcheries. *Trans. Amer. Fish. Soc.*, **108**, 323–34.

Ståhl G. (1983) Differences in the amount and distribution of genetic variation between natural populations and hatchery stocks of Atlantic salmon. *Aquaculture*, **33**, 23–32.

Stang P.M. & Seligman P.F. (1986) Distribution and fate of butyltin compounds in the sediment of San Diego Bay, California. In *Oceans '86 Conference Record*, Washington, DC, Sept. 23–25, 1986. 1256–61.

Stephenson R.L. (1990) Multiuse conflict—Aquaculture collides with traditional fisheries in Canada's Bay of Fundy. *World Aquaculture*, **21**(2), 34–45.

Stickney R.R. (1979) *Principles of warmwater aquaculture.* John Wiley & Sons, New York, 156–8.

Stickney R.R. (1988) Aquaculture on trial. *World Aquaculture*, **19**(3), 16–18.

Sullivan J.J., Simon M.G. & Iwoaka W.T. (1983) Comparison of HPLC and mouse bioassay methods for determining PSP toxins in shellfish. *J. Food Sci.*, **48**, 1312–4.

Sumari O. (1982) A report on fish farm effluents in Finland. In *Report of the EIFAC Workshop on Fish-farm Effluents* (Ed. by J.S. Alabaster). *EIFAC Tech. Pap.*, **41**, 21–7.

Sundli, A. (1999) Hommenkothen guidelines for sustainable aquaculture (adopted 1998), In *Sustainable Aquaculture—Good for the future.* (Eds. N. Svennevig *et al.*) Baltema, Rotterdam; 343–347.

Swingle H.S. (1969) *Methods of analysis for waters, organic matter and pond bottom soils used in fisheries research.* Auburn University, Auburn, Alabama, 119p.

Szluha A.T. (1974) Potamological effects of fish hatchery discharge. *Trans. Amer. Fish. Soc.*, **103**(2), 226–34.

Tacon A.G.J. (1996) Feeding tomorrow's fish: the Asian. In *Aquaculture '96* (Ed. by K.P.P. Nambiar & T. Singh) 20–42. Infofish, Kuala Lumpur.

Tacon A.G.J. & de Silva S.S. (1994) Feed preparation and feed management strategies within semi-intensive fish farming systems in the tropics. *Aquaculture* **151**, 379–404.

Tanabe S. *et al.* (1989) Isomer-specific determination and toxic evaluation of potentially hazardous coplanar PCBs, dibenzofurans and dioxins in the tissues of 'Yusho' PCB poisoning victim and in the causal oil. *Toxicol. Environ. Chem.* **24**, 215–31.

Tangen G. (1977) Blooms of *Gyrodinium aureolum* (Dynophyceae) in North European waters, accompanied by mortality in marine organisms. *Sarsia*, **63**(2), 123–33.

Taylor D. (1982) Minimata disease. *Environ. Sci. Technol.*, **16**, 81A.

Teas H.J. (1976) Productivity of Biscayne Bay mangroves. In *Biscayne Bay, Past, Present and Future* (Ed. by A. Thorhaug). *Univ. Miami Sea Grant Spec. Rep.*, **5**.

Teas H.J., Jurgens W. & Kimball M.C. (1976) Plantings of red mangroves (*Rhizophora mangle* L.) in Charlotte and St. Lucie counties, Florida. In *Proceedings of the Second Annual Conference on the Restoration of Coastal Vegetation in Florida* (Ed. by R.R. Lewis & D.P. Cole). Hillsborough Community College, Tampa, Florida.

Tenore K.R. *et al.* (1982) Coastal upwelling in Rias Bajas, N.W. Spain: contrasting the benthic regimes of the Rias de Arosa and de Muros. *Journal of Marine Research* **40**, 701–72.

Tenore K.R. *et al.* (1985) Effects of intense mussel culture on food chain patterns and production in coastal Galicia, N.W. Spain. In *Proceedings of the International Symposium on Utilization of Coastal Ecosystems: Planning, Pollution and Productivity, 21–27 Nov. 1982, Rio Grande, Brazil, Vol. 1* (Ed. by N.L. Chao and W. Kirby-Smith), 321–9.

Terchunian A. *et al.* (1986) Mangrove mapping in Ecuador: the impact of shrimp pond construction. *Environ. Mgmt.* **10**(3), 345–50.

Thain J.E. (1986) Toxicity of TBT to bivalves: Effects on reproduction, growth and survival. In *Ocean '86 Conference Record*, Washington, DC, Sept. 23–25, 1986, 1256–61.

Thompson P.E., Dill W.A. & Moore G. (1973) The major communicable fish diseases of Europe and North America—A review of national and international measures for their control. *EIFAC Tech. Pap.*, **17**(Suppl. 1), 48p.

Thorgaard G.H. & Allen S.K. (Jr.) (1988) Environmental impacts of inbred, hybrid and polyploid aquatic species. *J. Shellfish Res.*, **7**(3), 556.

Thorhaug A. (1987) Restoration of mangroves and seagrasses and attendant economic benefits for fisheries and mariculture; management, policy and planning. In *Papers contributed to the Workshop on Strategies for the Management of Fisheries and Aquaculture in Mangrove Ecosystems, Bangkok, Thailand, June 1986* (Ed. by R.H. Mepham). *FAO Fish. Rep.*, **370** Suppl., 142–60.

Thorpe J. (Ed.) (1980) *Salmon ranching*. Academic Press, London, 441p.

Tiews K. (Ed.) (1981) *Aquaculture in heated effluents and recirculation systems, Vol. 1*. Schriften der Bundesforschungsanstalt für Fischerei, Hamburg, 59–61.

Tilzey R.D.J. (1980) Introduced fish. In *An Ecological Basis for Water Resource Management* (Ed. by W.D. Williams). Australian National University Press, Canberra, 271–9.

Timmans M.B. & Lasordo T.M. (Eds) (1994) Aquaculture water-reuse systems: engineering design and management. *Developments in Aquaculture and Fisheries Science* **27**.

Tucholski S., Kok J. & Wojno T. (1980a) Studies on removal of wastes produced during cage rearing of rainbow trout (*Salmo gairdneri* Richardson) in lakes. 1. Chemical composition of wastes. *Rocz. Nauk. Rain.*, **82**, 3–15.

Tucholski S., Wieclawski F. & Wojno T. (1980*b*) Studies on removal of wastes produced during cage rearing of rainbow trout (*Salmo gairdneri* Richardson) in lakes. 2. Chemical composition of water and bottom sediments. *Rocz. Nauk. Roln.*, **82**, 17–30.

Tucholski S. and Wojno T. (1980) Studies on removal of wastes during cage rearing of rainbow trout (*Salmo gairdneri* Richardson) in lakes. 3. Budgets of mineral material and some nutrient elements. *Rocz. Nauk. Roln.*, **82**, 31–50.

Tucker J.B. (1983) Schistosomiasis and water projects: breaking the link. *Environment*, **25**(7), 17–20.

Turner H.J. (1949) *Report on investigations of methods of improving shellfish resources of Massachusetts.* Dept. of Conservation, Division of Marine Fisheries, Massachusetts, 22p.

Turner R.E. (1979) Louisiana's coastal fisheries and changing environmental conditions. In *Proceedings of the Third Coastal Marsh Estuary Symposium* (Ed. by J.W. Day *et al.*). Baton Rouge, Louisiana, Louisiana State University, 363–70.

Turner G.E. (Ed.) (1988) *Codes of practice and manual of procedures for consideration of introductions and transfers of marine and freshwater organisms. EIFAC Occas. Pap.*, **23**, 44p. and *ICES Coop. Res. Rep.*, **195**, 44p.

Twilley R.A., Lugo A.E. & Patterson-Zucca C. (1986) Litter production and turnover in basin mangrove forests in Southwestern Florida. *Ecology*, **67**, 670–83.

UNEP (1988) *Environmental impact assessment—Basic procedures for developing countries.* UNEP Regional Office for Asia and the Pacific, Bangkok, 16p.

UNEP (1990) *An approach to environmental impact assessment for projects affecting the coastal and marine environment.* UNDP Regional Seas Reports and Studies, No. 122. UNEP, Nairobi, 1–12.

Ungson J.R. *et al.* (1993) An economic assessment of the production and release of marine fish fingerlings for sea ranching. *Aquaculture* **18**, 169–81.

University of Stirling (Institute of Aquaculture) (1988) *The reduction of the impact of fish farming on natural environment.* University of Stirling, Stirling, **IV**, 167p.

University of Stirling (Institute of Aquaculture), Institute of Freshwater Ecology and Institute of Terrestrial Ecology (1990) *Fish farming and the Scottish freshwater environment*, Nature Conservancy Council, Edinburgh, Scotland, 285p.

US EPA (1999) Guidelines for carcinogen risk review. *Risk Assessment Forum, Washington, DC.* NCEA-F-0644, July 1999, Review Draft.

Van Houtte A.R., Bonucci N. & Edeson W.R. (1989) *A preliminary review of selected legislation governing aquaculture.* ADCP/REP/89/42. UNDP/FAO, Rome, 81p.

Van Vessem J., Draulans D. & de Boot A.F. (1985) The effects of killing and removal on the abundance of grey herons at fish farms. *XVII Congress of the International Union of Game Biologists, Brussels, Sept. 1985*, 337–43.

Villwock N. (1963) Die Gattung *Orestias* (Pisces, Microcyprini) und die Frage der intralakustrichen speziation in Titicascengebiet. *Verb. dt. Zool. Ges. Wien, Zool. Anz. Suppl.*, **26**, 610–24.

Vollenweider R.A. (1968) *Scientific fundamentals of the eutrophication of lakes and flowing waters, with particular reference to nitrogen and phosphorus as factors in eutrophication.* Report to OECD, Paris, DAS/CSI/68.

Vollenweider R.A. (Ed.) (1971) *A manual on methods for measuring primary production in aquatic environments. IBP Handbook No. 12.* Blackwell Scientific Publications for IBP, Oxford, 213p.

Vollenweider R.A. (1975) Input–output models with special reference to the phosphorus loading concept in limnology. *Schweiz. Z. Hydrol.* **37**, 53–84.

Walker C.R. (1961) Toxicological effects of herbicides on the fish environment (Part 1 and 2). *Water & Sewage Works*, **3**, March and April 1961.

Wallen I.E. (1951) The direct effect of turbidity on fishes. *Okla. Agric. Mech. Coll. Bull.*, **48**, 1–27.

Walsh G.E. (1967) An ecological study of a Hawaiian mangrove swamp. In *Estuaries* (Ed. by G.H. Lauff). Publication No. 83, American Association for the Advancement of Science, Washington, DC, 420–31.

Warrer-Hansen I. (1982) Methods of treatment of waste water from trout farming. In *Report of the EIFAC Workshop on Fish-Farm Effluents, Denmark, May 1981* (Ed. by J.S. Alabaster). *EIFAC Tech. Pap.*, **41**, 113–21.

Warrer-Hansen I. (1989) Production and control of wastes in freshwater fish farming. *Aquaculture Ireland*, **40**, 19–22.

Welch H.W. & Lewis G.D. (1976) Assessing environmental impacts of multiple use land management. *J. Environ. Management*, **4**, 197–209.

Welcomme R.L. (1984) International transfers of inland fish species. In *Distribution, Biology and Management of Exotic Fishes* (Ed. by W.R. Courtenay, Jr. & J.R. Stauffer, Jr.). Johns Hopkins University Press, Baltimore, 22–40.

Welcomme R.L. (1988) International introductions on inland aquatic species. *FAO Fisheries Technical Paper*, **294**, 318. Rome.

Welcomme R.L. & Bartley D.M. (1998) An evaluation of present techniques for the enhancement of fisheries. In *Inland Fishery Enhancements* (Ed. by T. Petr) 1–35. FAO Fisheries Technical Paper No. 374. FAO, Rome.

Weston D.P. (1986a) *The environmental effects of floating mariculture in Puget Sound.* Special Report, Washington Dept. of Fisheries and Ecology, Seattle, USA.

Weston D.P. (1986b) *Recommended interim guidelines for the management of salmon net-pen culture in Puget Sound.* Report prepared by Science Application International Corporation for Washington Department of Ecology in conjunction with the Departments of Fisheries, Aquaculture and Natural Resources.

Weston D.P. & Gowen R.J. (1988) *Assessment and prediction of the effects of salmon net-pen culture on the benthic environment.* Washington Department of Fisheries, Technical Report, 414p.

Whetstone G.A., Parker H.W. & Wells D.M. (1974) *Study of current and proposed practices in animal waste management.* EPA 430/9-74-003. United States Environmental Protection Agency, Washington, DC.

White A.W. (1987) *Blooms of toxic algae worldwide: their effects on fish farming and shellfish resources. Conference 2. The impact of toxic algae on mariculture.* AQUANOR 87, Trondheim, Norway, 9–14.

White A.W. et al. (1985) Inability of oxonation to detoxify shellfish toxins in the soft-shell clams. In *Toxic Dinoflagellates* (Ed. by D.M. Anderson, A.W. White & D.G. Baden). Elsevier, New York, 473–8.

WHO (1989) Health guidelines for the use of wastewater in agriculture and aquaculture. *WHO Technical Report Series*, **778**, 43p.

Wickins J.F. (1981) Water quality requirements for intensive aquaculture: A review. In *Aquaculture in Heated Effluents and Recirculation Systems, Vol. 1* (Ed. by C. Tiews). Schriften der Bundesforschungsanstalt für Fischerei, Hamburg, 17–37.

Wiesmann D., Scheid H. & Pfeffer E. (1988) Water pollution with phosphorus of dietary origin by intensively fed rainbow trout (*Salmo gairdneri* Rich.). *Aquaculture*, **69**, 263–70.

Willoughby H., Larsen H.N. & Bowen J.T. (1972) The pollutional effects of fish hatcheries. *Amer. Fishes & U.S. Trout News*, **17**(3), 6–7, 20–1.

Wilson E.O. (1965) The challenge from related species. In *The Genetics of Colonizing Species* (Ed. by H.G. Baker & G.E. Sleblins). Academic Press, London, 588p.

Wilson, R.P. (1994) Utilisation of dietary carbohydrate by fish. *Aquaculture* **124**, 67–80.

Woodiwiss F.S. (1964) A biological system of stream classification used by the Trent River Board. *Chem. Ind.*, 1964, 443–7.

Woodward I. (1989) Finfish farming and the environment: a review. *Tech. Rep. Dep. Sea Fish.*, Tasmania, **35**, 43p.

Woynarovich E. (1980) Raising ducks on fish ponds. In *Integrated Agriculture-Aquaculture Farming Systems* (Ed. by R.S.V. Pullin & Z.H. Shehadeh). *ICLARM Conf. Proc.*, **4**, 129–34.

Wray T. (1988) Floating raceway designed to cut sea farm risks. *Fish Farm. Int.*, **15**(7), 4–5.

Yamazaki F. (1983) Sex control and manipulation in fish. *Aquaculture*, **33**, 329–54.

Yentch C.M. & Ineze L.S. (1980) Accumulation of algal biotoxins in mussels. In *Mussel Culture in North America* (Ed. by R. Lutz). Elsevier, New York, 223–46.

Yentsch C.S. (1987) *Monitoring algal blooms, the use of satellites and other remote sensing devices. Conference 2, The impact of toxic algae on mariculture.* AQUANOR 87, Trondheim, Norway.

Zenny F.B. (1969) Comparative study of laws and regulations governing the international traffic in live fish and fish eggs. *EIFAC Tech. Pap.* **(10)**, 57p.

Zirschky J. & Reed S.C. (1988) The use of duckweed for wastewater treatment. *J. Water Pollut. Control Fed.*, **60**(7), 1253–8.

Zitko V. (1986) Chemical contamination in aquaculture. *Canadian Aquaculture*, **2**(1), 9–10.

Index

Lightning Source UK Ltd.
Milton Keynes UK
UKOW03n1018080515

251143UK00001B/4/P